Richard Jones Phillips

Spectacles and Eyeglasses

Their forms, mounting, and proper adjustment

Richard Jones Phillips

Spectacles and Eyeglasses
Their forms, mounting, and proper adjustment

ISBN/EAN: 9783337317102

Printed in Europe, USA, Canada, Australia, Japan

Cover: Foto ©berggeist007 / pixelio.de

More available books at **www.hansebooks.com**

SPECTACLES

AND

EYEGLASSES

THEIR FORMS MOUNTING AND
PROPER ADJUSTMENT

BY

R. J. PHILLIPS, M.D.,

INSTRUCTOR IN DISEASES OF THE EYE, PHILADELPHIA POLYCLINIC AND COLLEGE
FOR GRADUATES IN MEDICINE; OPHTHALMIC SURGEON TO THE
PRESBYTERIAN HOSPITAL IN PHILADELPHIA, ETC.

WITH 47 ILLUSTRATIONS.

PHILADELPHIA:
P. BLAKISTON, SON & CO.,
1012 WALNUT STREET.
1892.

PREFACE.

This little work is the outgrowth of the instruction on the subject of prescribing spectacle frames which has been given to successive classes at the Philadelphia Polyclinic and College for Graduates in Medicine. The book, like the teaching referred to, is intended to supplement studies in refraction, and to give the student that knowledge of the correct placing of the glasses before the eyes without which the most painstaking measurement of the refraction will frequently fail of practical result. With the popularization, as one may call it, of ophthalmology in the profession, many physicians who prescribe glasses are compelled, by the lack of skilled opticians in their neighborhood, to themselves furnish the spectacles to the patient. To these, it is believed, the knowledge which I have endeavored to impart in these pages will prove especially useful.

Of late years much advance has been made in the art of making efficient, comfortable and handsome contrivances for holding glasses before the eyes, and the increased use of prismatic and cylindrical lenses has given the fitting of the frames increased importance. Text-books of refraction remain, however, almost devoid of reference to the subject, the scant literature of which is scattered through opticians' trade publications, and a few medical periodicals. Free application has been made to such sources, and the indebtedness incurred duly acknowledged in the text.

My thanks are due to my friend and instructor, Dr. Edward Jackson, for many valuable suggestions in writing this treatise, and, indeed, for directing my attention to the need of a book on spectacles.

Dr. George M. Gould kindly furnished me with some references used in the introduction, and I am indebted to The Philadelphia Optical and Watch Co., and to Messrs. Wall & Ochs, for a number of cuts.

March, 1892.

CONTENTS.

	PAGE
INTRODUCTION,	17

I. GENERAL CONSIDERATIONS, 26
The Material of Frames, ... 26
The Component Parts of Spectacles, ... 27
The Lenses : Their Material and Manufacture, . . 28
Eye Wires, Temples, and Bridges, ... 34
The Different Patterns of Spectacles, . . . 34
The Varieties of Eyeglasses, ... 40
Spectacles for Cosmetic Effect, ... 43

II. THE PRINCIPLES OF SPECTACLE FITTING, ... 44
Centering and Decentering, ... 44
Prismatic Effect of Decentering, ... 46
Normal Lateral Centering, ... 49
Normal Vertical Centering, ... 50
Distance of the Glasses from the Eyes, ... 51
Perpendicularity of the Plane of the Lenses to the Visual Axes, . 52
Periscopic Glasses, ... 56

III. PRESCRIPTION OF FRAMES, ... 59
The Measurements Required, ... 59
Obtaining the Interpupillary Distance, ... 61
Height of the Bridge, ... 65
Relation of the Top of the Bridge to the Plane of the Lenses, . . 66
Width of Base, ... 67
Prescription of Eyeglasses, ... 69

IV. INSPECTION AND ADJUSTMENT OF SPECTACLES AND EYEGLASSES, 70
Proving the Strength of Lenses, ... 70
Phacometers, ... 71
Neutralization of Spherical Lenses, ... 73
Neutralization of Cylindrical Lenses, ... 74

CONTENTS.

INSPECTION AND ADJUSTMENT OF EYEGLASSES, ETC.—*Continued.* PAGE

Neutralization of Sphero-cylindrical Lenses, ... 75
Locating the Optical Center, ... 76
Finding the Apex of a Prism, ... 77
Measuring the Strength of a Prism, ... 77
Detection of Scratches, Specks, Flaws, Etc., ... 82
Irregularity of the Refracting Surfaces, ... 83
Adjusting Spectacle Frames, ... 83
Adjustment of Eyeglasses, ... 89
The Care of Spectacles, ... 92

LIST OF ILLUSTRATIONS.

FIGURE		PAGE
1.	Position of the parts of spectacles,	28
2.	Position of the parts of eyeglasses,	28
3.	Optician's lens-grinding lathe,	30
4.	Concave toric and concave cylindrical lens,	33
5.	Frameless bifocal spectacles,	35
6.	Forms of spectacle bridges,	36
7.	Ovals, showing the sizes of spectacle eyes,	38
8.	Forms of bifocal glasses,	39
9.	Extra front,	41
10.	Forms of rigid frame eyeglasses,	41
11.	Modern frameless eyeglasses,	42
12.	Spectacles with lenses decentered "in,"	45
13.	Section of a normally centered lens,	47
14.	Decentered lens, showing prismatic effect,	47
15.	Profile view of the face, showing the "natural" position for the spectacle bridge,	51
16.	Top of bridge "out" from plane of the lenses,	52
17.	Top of bridge "in" from plane of the lenses,	52
18.	Spectacles facing directly forward,	54
19.	Spectacles facing downward and forward,	54
20.	Spectacles facing inward,	55
21.	Front and back of a convenient spectacle rule,	60
22.	Method of measuring the height of a spectacle bridge,	62
23.	Simplest method of measuring the interpupillary distance,	62
24.	Maddox pupil localizer,	63
25.	Method of using the Maddox pupil localizer,	63
26.	Meyrowitz pupilometer,	64
27.	A common form of pupilometer,	65
28.	Method of measuring the distance of the bridge "out,"	66
29.	Method of measuring the distance of the bridge "in,"	67
30.	Mr. Brayton's lens measure,	72
31.	Apparent displacement of lines caused by rotating a cylinder,	74

LIST OF ILLUSTRATIONS.

FIGURE	PAGE
32. Method of finding the axis of a cylindrical lens,	74
33. Ready method of locating the optical center of a lens,	76
34. Mode of marking the apex of a prism,	77
35. A prism improperly held,	77
36. Implement for measuring the refracting angle of prisms,	80
37. Method of determining the deviating angle of prisms,	81
38. Rotation of a lens within its eyewire,	83
39. Bend at the junction of the eyewire and bridge,	84
40. Showing the planes of the lenses crossing each other,	84
41. Bend of the bridge,	85
42. Inequality of corresponding angles of the bridge,	85
43. Angles on one side of the bridge too small; on the other too large,	85
44. Proper fitting of hook temples,	88
45. A common but incorrect shape of the nose pieces of eyeglasses,	89
46. Eyeglasses with nose pieces of correct shape,	90
47. The points at which eyeglasses are to be adjusted,	91

SPECTACLES AND EYEGLASSES.

INTRODUCTION.

At what time man invented lenses and discovered the aid which they are capable of lending to vision is a matter beyond our knowledge. It is tolerably certain that they were known to civilizations earlier than ours. The late Wendell Phillips was wont to assert that spectacles were among the things known to the ancients. Though it might be difficult to sustain this assertion as regards spectacles in the present meaning of the term, the evidence in relation to their acquaintance with the essential element of spectacles, the lens, is reasonably convincing. This evidence was for the most part discovered by Sir Austen Henry Layard, among the ruins of old Nineveh, and is of the most interesting character. Among the articles which he unearthed, was a specimen of *transparent* glass (a small vase or bowl) with a cuneiform inscription fixing its date quite accurately to the latter part of the seventh century B. C. ("Discoveries Among the Ruins of Nineveh and Babylon, etc.," by Austen H. Layard, New York, 1853, p. 196.) This is the most ancient known specimen of transparent glass, though Egypt furnishes it of a date only a century later, and opaque or colored glass was manufactured at a much earlier period; some specimens of the fifteenth century B. C. still enduring. However, the ancient nations were not compelled to wait for transparent glass in order to invent lenses, as they had in rock crystal a material admirably adapted to that purpose, and Layard was so fortunate as to discover such a lens in Nineveh. (*Ibid.* p. 197). Sir David Brewster, who examined this lens, described it as being plano-convex, of a diameter of one and a half inches, and capable of forming a tolerably distinct focus at a distance of four and a half inches from the plane side. It is interesting to note farther in regard to this, the oldest lens in existence, that it is fairly well polished, though somewhat uneven from the mode in which it was ground, which Brewster concludes was not upon a spherical surface, but by means of a lapidary's wheel, or some method equally rude. Another evidence of the use of lenses has come down to us from antiquity. Upon record-cylinders of old Nineveh, and on engraved gems and stones of Babylon, Egypt, and other sources which long antedate the Christian era, are characters and lines of such delicacy and minuteness as to be undecipherable without the aid of a magnifying lens. Taking these facts in conjunction, the statement that some of the properties of lenses were known to and utilized by the ancients, the old record writers of Assyria, for instance, may be regarded as almost as well demonstrated as though it were made of a modern engraver, and we were to step into his workshop and find his magnifying loup lying beside his work.

The testimony as to their use by the Romans during their supremacy is of a less conclusive character. The statement frequently made that the Emperor Nero used a concave jewel to assist his sight rests upon some obscure sentences in Pliny. That author says: "Nero could see nothing distinctly without winking and having it brought close to his eyes." (Bk. 11, Chap. 54, Riley's Trans.) In another place, speaking of the emerald, *smaragdus*, he says: "In form these are mostly concave, so as to reunite the rays of light and the powers of vision. * * * When the surface of the smaragdus is flat, it reflects the image of objects in the same manner as a mirror. The Emperor Nero used to view the combats of gladiators upon (with, or by means of) a smaragdus." (Bk. 37, Chap. 17.) The mention of the reflecting properties of the emerald immediately before the statement of Nero's use of it, with the alternative renderings of the Latin ablative, *smaragdo*, make the supposition that Nero used the emerald as an eyeglass uncertain, though in view of his clearly described nearsightedness, the conjecture is probable enough.

Lenses appear to have been unknown in Europe during the first twelve hundred years of the Christian era, though the Saracen Alhazen, who died in Cairo in 1038, has left books showing his acquaintance with them. These books were brought to Europe at a very early period, and the manuscripts yet exist; some in the Bodleyan library, and another portion in that of the University of Leyden. It was probably from them that the early writers obtained their first hints of the science of optics, on the revival of learning in the fourteenth and fifteenth centuries. It is worthy of note that Alhazen was born at Bassora, at the head of the Persian Gulf, and less than five hundred miles from the spot where, sixteen hundred years before, had stood the palace of the Assyrian kings in the ruins of which Sir Henry Layard found the lens of crystal. It might, perhaps, be plausibly maintained that in the countries about the Tigris some knowledge of optics, and of convex lenses, has persisted without eclipse from the most remote ages.

The earliest European reference to our subject occurs in the writings of Roger Bacon, who died in 1292, and to whom the invention of the instrument he describes is sometimes accredited. Bacon's glass was apparently a large plano-convex lens, probably what we now call a reading glass, intended to be held in the hand, and of it he says: "This instrument is useful to old men and to those that have weak eyes; for they may see the smallest letters sufficiently magnified." Spectacles proper; that is, glasses mounted so as to retain themselves upon the face, appear to have been invented in Florence somewhere between 1280 and the close of the thirteenth century. Dr. Samuel Johnson is said to have expressed surprise that the inventor of such useful articles has found no biographer. Doubtless among the thousands for whom the discovery has kept open the sources of knowledge there would be found one to pay this tribute to the fame of his benefactor were the identity of the latter a matter of certainty. But, unfortunately, our evidence on the point is of the most fragmentary character. We are told in a general way that the Chinese have for ages employed spectacles for the relief of defective eyesight. This is, perhaps, to be regarded as only another instance of the exercise of that claim to priority which the Chinese are known to extend over every good and perfect gift. The longest chase signally fails to bring the tradition to bay in any fact. The tomb of Salvinus Armatus, a Florentine nobleman who died in 1317, is said to bear an inscription to the effect that he was the inventor. If epitaphs enjoyed a less equivocal reputation for truthfulness he would doubtless be held in grateful remembrance as the man who has lengthened youth by postponing old age; and, like Joshua, kept back the night until the day's work was done.

Whoever the inventor, Alessandro di Spina, a monk of Florence who died in 1313, is generally accredited with having made public the use of spectacles, and by several Florentine writers of that time we find them mentioned and recommended. Pissazzo, in a manuscript written in 1299, says: "I find myself so pressed by age that I can neither read nor write without those glasses they call spectacles, lately invented, to the great advantage of poor old men when their sight grows weak." Friar Jordan, of Pisa, in 1305 says that "it is not twenty years since the art of making spectacles was found out, and is indeed one of the best and most necessary inventions in the world."

An early mention of spectacles, or in the language of that time, "a spectacle," occurs in "The Canterbury Tales," where Chaucer makes the Wife of Bath use the metaphor:—

> Povert (poverty) full often when a man is lowe,
> Makith him his God and eek himself to knowe.
> Povert *a spectacle* is, as thinkith me,
> Through which he may his verray frendes se.

There is in existence in the church of Ogni Santi, Florence, an old fresco by Domenico Ghirlandajo, representing St. Jerome, and dated 1480. The saint is portrayed seated at a desk, apparently deep in the composition of one of the blasts against the Heretics for which he was famous. Upon a peg at the side of the desk, together with the ink horn and a pair of scissors, hangs a small handleless *pince-nez*. The glasses are round and framed in dark bone, and in the bridge, also of bone, is a hinge. Though the artist seems to have been little impressed by the fact that St. Jerome died in the year 420, nearly nine centuries before spectacles were invented, the mounting and material represented in these early spectacles are worthy of note as showing their form in Ghirlandajo's time, and probably that in which they originated.

In the early references to spectacles it is the convex lens

for the use of the presbyopic which is mentioned. Concave lenses were probably introduced soon afterward; by whom we do not know. Glasses were at that time and for long afterward selected and used empirically; since it was not until the year 1600 that the astronomer, Johann Kepler, who may be regarded as the father of ophthalmology, made known in what manner the rays of light were refracted by the media of the eye and form an image upon the retina. Kepler went farther, and showed how convex and concave glasses influence this refraction, and to him is therefore due the honor of first scientifically treating this subject.

It must have been early discovered that there is a more or less close relation between the age of the wearer and the strength of the convex glass required, and the baneful theory was soon developed that this relation is constant, and that it would be ruinous to use a lens "too old for the eyes;" a superstition from which the public is even yet not fully emancipated. We find it rampant in Pepys' time, preventing his oculist, Dr. Turberville,* from giving that gentleman a proper correction for his accommodative asthenopia, of which the diary gives an accurate picture, and losing to the world many a priceless page. Pepys says (June 30, 1668): "My eyes bad, but not worse, only weary with working. * * * I am come that I am not able to read out a small letter, and yet my sight good, for the little while I can read, as ever it was, I think." But Dr. Turberville warns him against glasses too old for him, and so the diary is closed, and Pepys in a last pathetic entry resigns himself to coming blindness; and yet the

---

* Daubigny Turberville; created M.D. at Oxford in 1660. He practiced with great reputation as an oculist in London. His monument yet remains in Salisbury Cathedral, where he was buried.

convex lenses were at his hand, ready to dissipate the mists before him and enable him to "gaze upon a renovated world."

Improvement in spectacles appears to have been slow. The world waited more than two centuries after Kepler for another signal advance. Sir David Brewster is said to have discovered his own astigmatism; that is, he discovered that vertical and horizontal lines were not equally well seen by him at like distances, but the phenomenon was not explained and the observation faded from view. It remained for George Airy, the astronomer, to rediscover astigmatism, which he did about 1827, to determine that the curvature of the cornea was greater in one diameter than in another at right angles to the first, and to invent the cylindrical lens for the correction of the condition. Mr. Airy's right eye was myopic, while in the left he had compound myopic astigmatism. By a careful comparison of the appearance of objects when viewed with each eye singly, and a study of the effect of concave lenses held before the left eye upon lines crossing each other at right angles, he was able to conclude that the refraction of that eye differed in different planes. Mr. Fuller, an optician of Ipswich, made, under Airy's direction, a concave spherocylindrical lens which satisfactorily corrected his refractive error. Thus was the last great discovery in spectacles accomplished; a bit of work for completeness leaving nothing to be desired, and of not sufficiently acknowledged importance to humanity.

Benjamin Franklin invented bifocal spectacles. Since this statement is supposed by many to rest on tradition only, it may be of interest to quote a portion of a letter of Franklin's which bears upon the point. The letter is addressed to George Whately, of London, and is dated

Passy, 23d May, 1785. In it Dr. Franklin says: "By Mr. Dolland's saying that my double spectacles can only serve particular eyes, I doubt he has not been rightly informed of their construction. I imagine it will be found pretty generally true that the same convexity of glass through which a man sees clearest and best at the distance proper for reading, is not the best for greater distances. I therefore had formerly two pairs of spectacles which I shifted occasionally, as in traveling I sometimes read, and often wanted to regard the prospects. Finding this change troublesome and not always sufficiently ready, I had the glasses cut and half of each kind associated in the same circle. By this means, as I wear my spectacles constantly, I have only to move my eyes up or down, as I want to see distinctly far or near, the proper glasses being always ready. This I find more particularly convenient since my being in France. * * *" ("The Complete Works of Benjamin Franklin." Ed. by John Bigelow, New York, 1888.)

We may infer from the context that the invention took place before Franklin went to France, which was in the latter part of 1776. As he was born in 1706, the necessity for a double glass would first arise about 1750, and the invention therefore took place somewhere between this date and that of the journey to France.

The frames in which spectacles were mounted continued to be very clumsy affairs until the beginning of this century, when light metal frames were introduced in place of the earlier devices of bone, horn, or shell. Their later evolution has generally been along the lines of improved mechanical construction, and increased lightness and beauty. It would be difficult to mention an article which plays a more important part in modern life than do spectacles, or one which plays its part more acceptably. It is scarcely possible to estimate them at their true worth, or to imagine our condition without them. Deprived of their aid, most men would be too old for work at fifty, and purblind at sixty. For us all, as an old writer quaintly observes, "they keep the curtain from falling until the play has come to an end."

I. GENERAL CONSIDERATIONS.

By far the most generally useful method of placing glasses before the eyes is by spectacle frames, though the eyeglass, or *pince-nez*, has advantages in some cases, from the facility and quickness with which it may be placed in position or removed. The superiority of eyeglasses in appearance and becomingness is another point not unworthy of consideration, as the glasses will surely be more constantly worn if they are becoming than if they are not so. Moreover, the patient is justly entitled to the correction of his refractive error with as little injury to his appearance as possible. The disadvantages of eyeglasses are, that for constant wear they are seldom so comfortable as spectacles; that on some faces it is nearly impossible to keep them in place; while, where the contained glass is cylindrical or prismatic, the rotary motion which it is possible for the glass to take is a serious, and sometimes fatal objection to their adoption.

Lorgnettes and single eyeglasses, or quizzing glasses, as they are called, are little more than playthings; though sometimes, as in aphakia, or high myopia, a strong convex or concave lens in one of these forms is of use when the spectacles constantly worn do not give the vision which may occasionally be required.

The material of spectacle frames is usually gold, silver, or steel. Various alloys have also been employed, and sold as aluminium, or nickel. So far as I have examined them, they consist principally of tin, and contain little or none of the metals whose names they borrow. Real nickel is too flexible a metal to be used with advantage for spectacle frames, while, so far, no means have been found of soldering aluminium firmly. Were this difficulty overcome, the lightness, stiffness and freedom from rust of aluminium would make it an excellent material for cheap frames. Silver, like nickel, is too flexible, except for workmen's protective goggles, or some such purpose, where very heavy frames are allowable. Gold, of from 10 to 14 karat, is, by far, the best material for frames. Finer than this it is too flexible, while if less pure it may blacken the skin. In the end, such frames are cheaper than steel, as, owing to the liability of the latter metal to rust when in contact with the moist skin, the gold will outlast it many times over. In eyeglasses, however, the parts are heavier, and the metal is not in contact with the skin; so that there is not the same liability to rust. The gold frames furnished by opticians in this country usually have a stamp mark on the inner side of the right temple, near the hinge, which denotes the fineness of the gold: thus 8 karat is marked $+$; 10 karat, θ; 12 karat, $*$; while 14 karat, or finer, is marked 14k, etc.

The Component Parts of Spectacles.—A pair of spectacles is made up of fifteen or seventeen pieces, whose positions are shown in Fig. 1. They are: two lenses, two eye wires, four end pieces, two screws, two pins, or dowels, two temples and one bridge. Sometimes the rings upon the temples, through which the dowels pass, are formed as separate pieces. Fig. 2 shows the name and position of each part of an eyeglass. A glance at the more important of the many interesting processes required in making these different parts will contribute to an understanding of the subject.

The Lenses.—The word lens is the Latin name of the lentil, a small bean. The resemblance in shape caused the

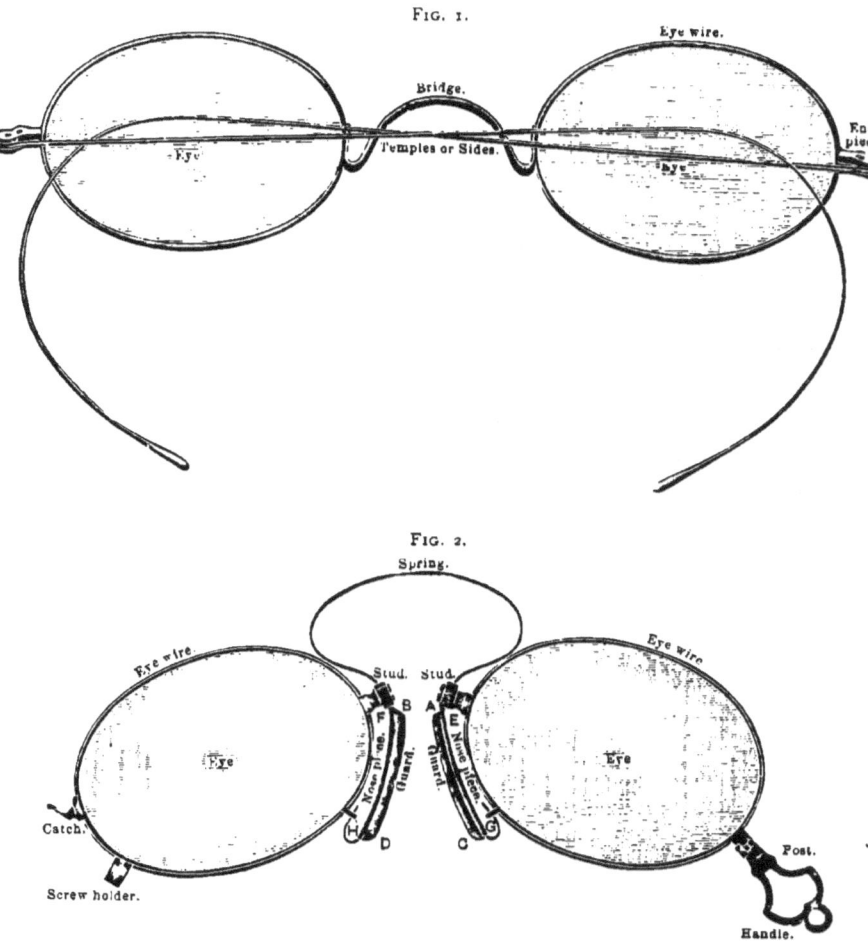

name to be applied to the optical implement. Spectacle lenses are usually made of glass; sometimes of rock crystal (crystallized quartz). The latter substance has a slightly higher index of refraction, so that a lens of a given strength may be somewhat lighter when made of it than when made of glass. The notion is common that these "pebbles," as they are called, possess a peculiar virtue in strengthening the eyes, or in some other direction. I suppose the idea is, that being the product of Nature's laboratory, they are necessarily superior. The advantage which they may have of being slightly lighter and harder than lenses of glass, is more than counterbalanced by their higher cost, and by the fact that the index of refraction of rock crystal is not very constant.

Of the different kinds of glass, that known as crown glass is preferred, on account of the superior brilliancy which it possesses. It differs from ordinary sheet glass only in the method of blowing. At one point in the process, the mass of glass on the blowpipe assumes the shape of a crown; hence the name. Although glass is theoretically a definite chemical compound, the different methods of handling make a considerable difference in the product of different makers. It consists, chemically, of silicic acid united with some two of the metallic bases: sodium, potassium, calcium, magnesium, aluminium, iron and lead, but owing to impurities in the glassmaker's raw materials, traces of several more of these bases are generally present. The bases calcium and soda are those used for ordinary sheet and crown glass; iron, always present as an impurity, giving the product its greenish tinge. To lessen this tint, arsenic is employed as a bleaching agent. Peroxide of manganese is sometimes used for the same purpose, and it is a slight excess of this agent which gives to certain samples of glass their pinkish tint. The transparency of

30   SPECTACLES AND EYEGLASSES.

such glass is thought to be less durable than that having the greenish color.

The simple apparatus used for grinding a single spherical lens is shown in Fig. 3. The disc of glass (*a*) of which a lens is to be made is fastened, by means of pitch, to a small, cubical block of iron (*b*) having a pit in the surface opposite that to which the glass is fastened. Into this pit fits a

FIG. 3.

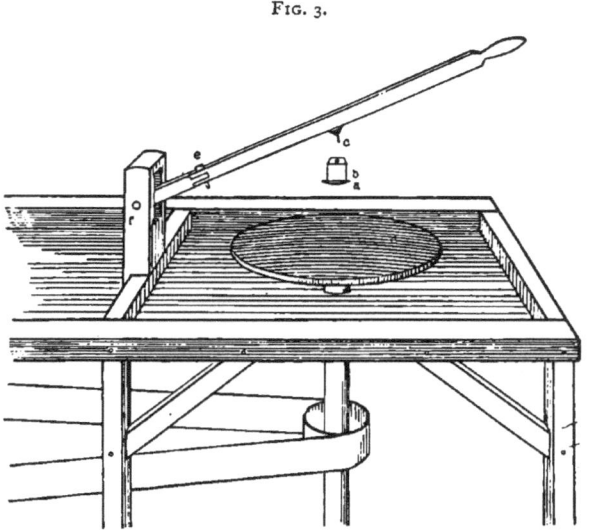

OPTICIAN'S LATHE, FOR GRINDING SPHERICAL LENSES.

pin (*c*) upon a lever which is in the hand of the workman. When the free surface of the glass is applied to the surface of the "tool," (*d*) to whose form it is to be ground, it, together with the block of iron, turns upon the pin. The joints at *e* and *f* allow of lateral and vertical movements of the lever, so that the workman is able to carry the glass freely over all portions of the tool.

GENERAL CONSIDERATIONS. 31

The tool which gives the shape to the surface of the glass is made of steel; and for spherical glasses is in the form of a disc, with its surfaces looking upward and downward, and revolving about a vertical axis, like a potter's wheel. The upper surface of this disc is convex for grinding concave glasses, or concave for grinding convex glasses. Of course each strength of lens requires a separate tool having the requisite convexity or concavity of surface. The abrading material placed upon the surface of the tool is wet powdered emery, of successively finer and finer grades until the desired amount of glass has been ground away. When this process is complete, the surface of the glass has the desired spherical curvature but it is rough: that is, it is "ground glass." To polish it a piece of wet broadcloth or felt is smoothly applied to the surface of the tool upon which the glass was ground, conforming, of course, to that surface. The cloth being sprinkled with wet "rouge" (a carefully calcined sulphate of iron), gives the glass held against it a beautiful polish without altering its spherical curvature. The same processes must now be gone over with the other surface of the lens, after which it is cleaned and cut with a diamond to a shape suitable for its future mounting, and its edges dressed.

In grinding a cylindrical lens the surface of the tool is, of course, a portion of the surface of a cylinder; and the glass is ground by a to-and-fro motion. It is evident that the axis of the cylinder in the future spectacle need not be taken into account in grinding, but only in the process of cutting to shape for mounting. As a matter of fact, very few cylinders are, at present, ground in this country: the glass is brought from Europe with a cylinder already ground upon one side and glued to its block of iron for the grinding of a spherical or plane surface upon the other.

When the lenses are of high power it is of advantage that they be made in the form of a meniscus, giving what are known as periscopic glasses. For instance, if a + 4. diopter lens is required, the anterior surface is ground to a + 6. D. and the posterior surfaces to a — 2. D. It is just as advantageous to a cylindrical or sphero-cylindrical glass to be periscopic as it is to a spherical, but, under present methods of grinding, it is manifestly impossible to give them this form, as the cylinder is ground on one side, and the other ground to a plane or sphere as the case may be. Glasses which overcome the difficulty have, however, been made, and were described by Dr. George C. Harlan at the meeting of the American Ophthalmological Society in 1885, and again in 1889. From the latter communication I quote the following description of the glass:—

"The lens to which I wish again to call the attention of the Society consists of crossed cylinders ground on one surface of the glass, the other being left for any desired spherical curve. In this way a meniscus may be produced. Here, for instance, is a combination lens giving the effect of + 4. ◯ + 2. Cyl. To produce this effect crossed cylinders of + 4. and + 6. are required, supposing the other surface of the glass to be left plain. If we wish to give the periscopic form to this glass, it can be done by making the cylinders 6 and 8, and grinding a — 2. sphere on the other side. If a simple cylinder is needed, the spherical curve must equal that of the weaker cylinder.

"I learn from a recent publication by Dr. George J. Bull, of Paris, entitled 'Lunettes et Pince-nez,' that glasses similar to these have been made with more or less success before, but have never come into general use. Dr. Bull describes them under the name of '*verres toriques.*' The tore (Latin *torus*) is the surface engendered by a circle which turns about an axis situated on the plane of the circle. A familiar example of the torus is the circular convex moulding at the base of an architectural column. A glass ground upon a wheel having this form will present two cylindrical curves, at right angles to each other, one depending on the radius of the wheel, and the other on the

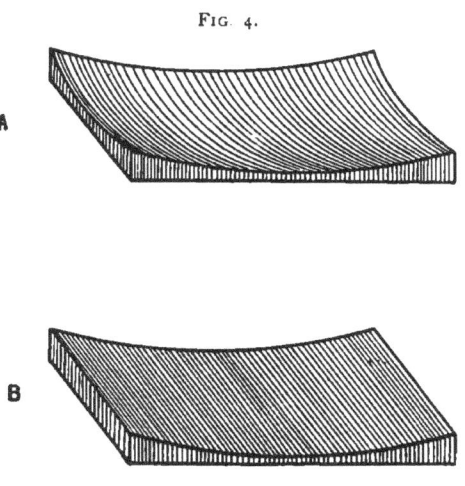

FIG. 4.

radius of the convexity of its rim. It would seem that 'toric lenses' is the proper designation of these glasses." Fig. 4, A, represents a concave toric lens. In the same figure, B is a concave cylindrical lens, introduced for the sake of comparison.

Those who have used these glasses consider them much more satisfactory than glasses made by the common method, and they should be borne in mind when prescribing for high astigmatism in patients who use their eyes a great deal for work requiring accuracy.

Eye Wires, Temples and Bridges.—Eye wires are made by wrapping the untempered wire, in the form of a spiral, closely about a flattened metal cylinder. Being tempered while in this position, the loops of the spiral will retain the shape given them. A single cut down the side of the cylinder converts each loop into a separate oval ring. End pieces and straight temples are stamped from sheets of metal, and afterward formed and tempered. Hook temples of steel are turned from wire upon a lathe. Bridges are usually made of oval or half oval wire, and are simply pressed to the desired shape by a forming machine.

Of the Different Patterns of Spectacles.—In the common and strongest form of spectacle, the edge of the glass is bevelled so as to enter a groove in the wire which surrounds it. In a second form, in which the edge of the glass is grooved for the reception of a fine, round wire, the object sought, of rendering the rim of the spectacles less conspicuous, is generally defeated by the fact that the glass must be made thicker than it otherwise need be, in order to give room for the groove on its edge. In concave glasses this is not the case, since the edge of the glass is here the thickest part, and such glasses may sometimes be mounted in this way with advantage. In a third form, called "frameless" spectacles (Fig. 5), the wire encircling the glass is dispensed with altogether, small holes being drilled through the glass near its edge for the accommodation of screws which fasten the bridge and temples in place. The advantages of this form are its beauty and inconspicuousness. It should never be prescribed for children, as it is quite liable to break at the point where the glass has been drilled. The edges of these glasses should not be polished, but should be given a dull finish, otherwise they reflect the light disagreeably.

Sides, or temples, have been variously constructed. Those having sliding and turn-pin joints are examples of antiquated forms. Those now used are the "hook," or "riding-bow," and the plain, "straight" temple. The former are to be preferred in all cases where the glasses are

FIG. 5.

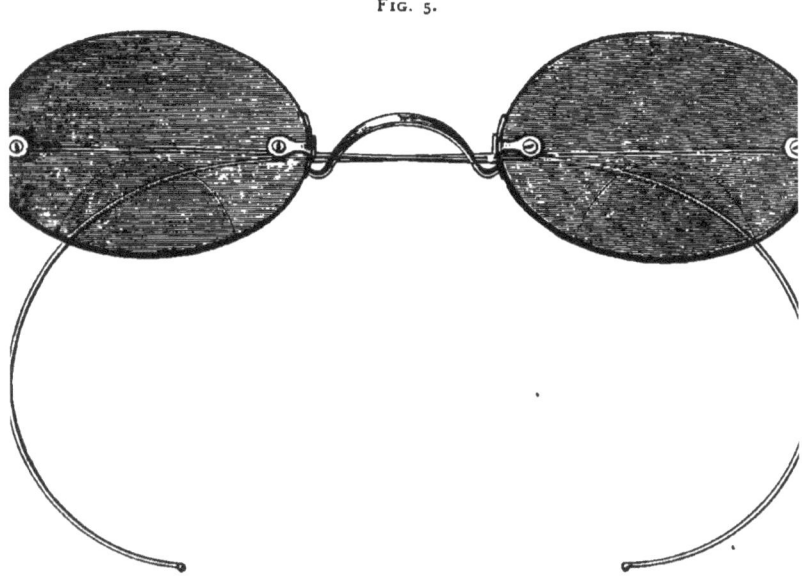

to be worn constantly or nearly so, and the latter for those who wear glasses for near work only, and require to remove them frequently from the eyes. Hook temples are made in three lengths, designated as short, medium, and long. These are sufficient for all cases.

Securing a proper fit in the bridge, upon which so much of the comfort and efficiency of spectacles depends, was a difficult matter until the ingenuity of Dr. Charles Hermon

Thomas, of this city, suggested what is known as the "saddle bridge," which solved the problem. (See Fig. 6.) This bridge may be varied to suit every possible case, and is always to be preferred. The "K" bridge, formed of wires in the shape of the letter K, is allowable in some cases. The nearly similar "X" bridge allows the glasses to teeter, or see-saw across the nose, with the motions of the head. The old-fashioned bridge called the "curl," is unobjectionable for cases in which the bridge of the nose

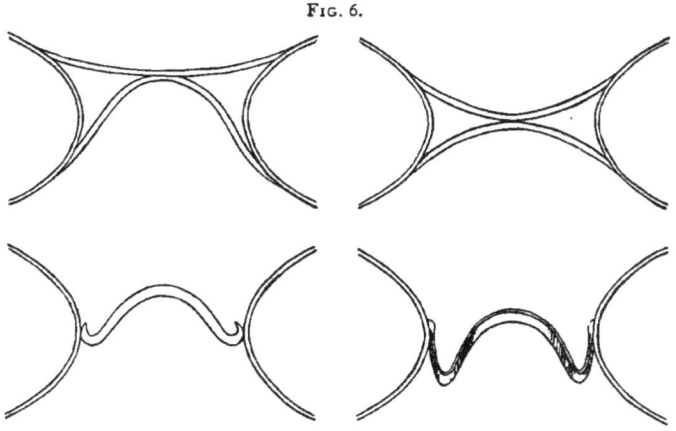

FIG. 6.

"X," "K," "CURL," AND "SADDLE" BRIDGES.

is prominent, or for the spectacles of old people, who like to slip their glasses down toward the end of the nose. None of the forms mentioned, however, have any advantage over the saddle bridge for any case. A small piece of cork is sometimes attached to the under side of the bridge where it comes in contact with the skin. It is unnecessary if the frames fit the face of the wearer properly. If it be desirable to remove all pressure from the bridge of the nose and to transfer it to the sides, it is best done by soldering a pair of guards, similar to those used on eyeglasses, to the spectacle bridge.

The earliest spectacles appear to have had round eyes. Various other shapes are occasionally seen, as octagon, oblong, etc. The oval has about displaced these antiquated forms, and is made in sizes known to the trade in America as follows:—

TABLE I—SIZES OF EYES.

No. 00. $1\frac{19}{32}$ by $1\frac{1}{4}$ in. or 41 by 32 mm. large coq. size.
" 0. $1\frac{17}{32}$ by $1\frac{13}{18}$ in. or 39 by 30 mm coq. size.
" 1. $1\frac{7}{10}$ by $1\frac{1}{8}$ in. or 37 by 28 mm. standard large eye.
" 2. $1\frac{13}{32}$ by $\frac{31}{8}$ in. or 36 by 25 mm. standard E. G. size.
" 3. $1\frac{3}{8}$ by $1\frac{1}{32}$ in. or 35 by 26 mm. standard interchange.
" 4. $1\frac{11}{12}$ by $\frac{31}{2}$ in. or 34 by 25 mm. standard small eye.
" 5. $1\frac{1}{4}$ by $\frac{29}{32}$ in. or 32 by 23 mm. children's size.

Where glasses are used for near work only, the eyes are sometimes made of semi-oval shape, allowing the line of sight to pass over their upper, straight edge when the wearer views a distant object. These are known as "half," "pulpit," or "clerical" eyes, and are very convenient, especially to public speakers, as their name implies. They do not seem to me as well known or as generally used as they should be.

When glasses of different focussing power are required for distant and near vision, the trouble incident to frequent changing is obviated by "bifocal" glasses. That is, the lower part of the spectacle eye, which is used for near work, is made to differ in focussing power from the upper part, which is used for distant vision. Such bifocal glasses are also called Franklin glasses, from the philosopher who, as we have seen, invented them.

The object sought may be attained in various ways. In the early Franklin glasses each eye contained two half oval

FIG. 7.

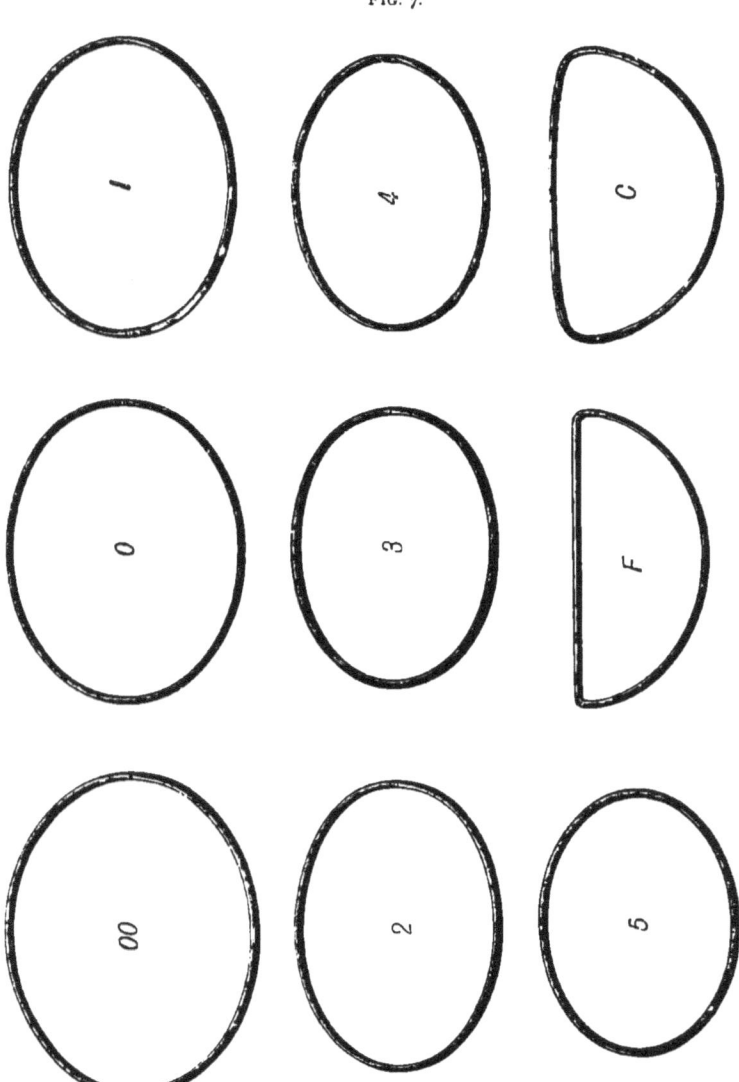

OVALS SHOWING THE ACTUAL SIZES OF EYES ACCORDING TO TABLE I.

pieces, with their straight edges in apposition (A, Fig. 8). This has been improved upon by making the line of junction a curved one, giving somewhat greater latitude of distant vision and rendering the glass more secure in its frame. A form of bifocal glasses which were never used to any great extent are called " ground " bifocals. They are very handsome, containing only one piece of glass in each eye, the upper and lower parts of which are ground of different strengths (C, Fig. 8). The mechanical impossibility

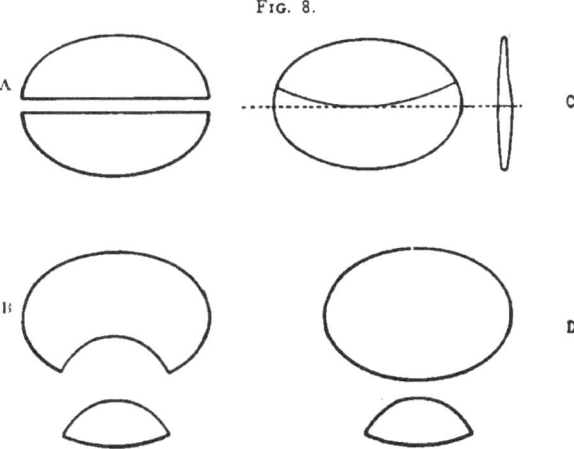

FIG. 8.

of centering both spherical surfaces upon the same piece of glass, however, introduces a prismatic effect which destroys their usefulness. The latest and best variety of these glasses are called "cemented bifocals" (D, Fig. 8). They have been in occasional use in France for over twenty years, though their general manufacture by our opticians is due to the efforts of Dr. Geo. M. Gould. To the back of the distance glass is cemented, by means of Canada balsam, a small lens whose strength added to that of the distance glass equals the strengh required for near work. The upper edge of the supplemental lens should be ground as thin as possible, in order to render it inconspicuous. This nearly does away with the objectionable line of junction, the spectacles are strong, light and handsome, and may even be made "frameless," like those represented in Fig. 5, if the patient so desire. For cylindrical lenses this form is also cheaper, since only the distance glass need have the cylinder ground upon it, the supplemental segment being a simple sphere.

All bifocals have the inconvenience that in walking, the floor just in front of the patient's feet is not seen clearly because viewed through the near glass. Spectacles which revolve on the long axis of the "eye," bringing the distance glass to the lower portion of the frame, have been contrived to overcome this difficulty, but they are cumbersome and, moreover, it requires more effort to effect the revolution of the glass than it does to bend the neck sufficiently to bring the upper segment into the line of vision when the ordinary bifocals are worn. Some persons declare that they cannot become accustomed to bifocals however well adjusted. Parallel, horizontal lines, as those of a staircase, are particularly confusing, it being possible to see each line doubled if the junction of the two segments of the glass is placed just opposite the pupil. Such persons may prefer having an "extra front" (Fig. 9): that is, a second pair of spectacles whose temples are replaced by short hooks, by means of which they are hung in front of the frame already upon the face. This is a rather clumsy device; less so, however, when the eyes of the extra front are made half oval instead of oval.

Eyeglasses.—Most that has been said of the varieties of spectacle frames applies as well to eyeglass frames. As

GENERAL CONSIDERATIONS. 41

Fig. 9.

Fig. 10.

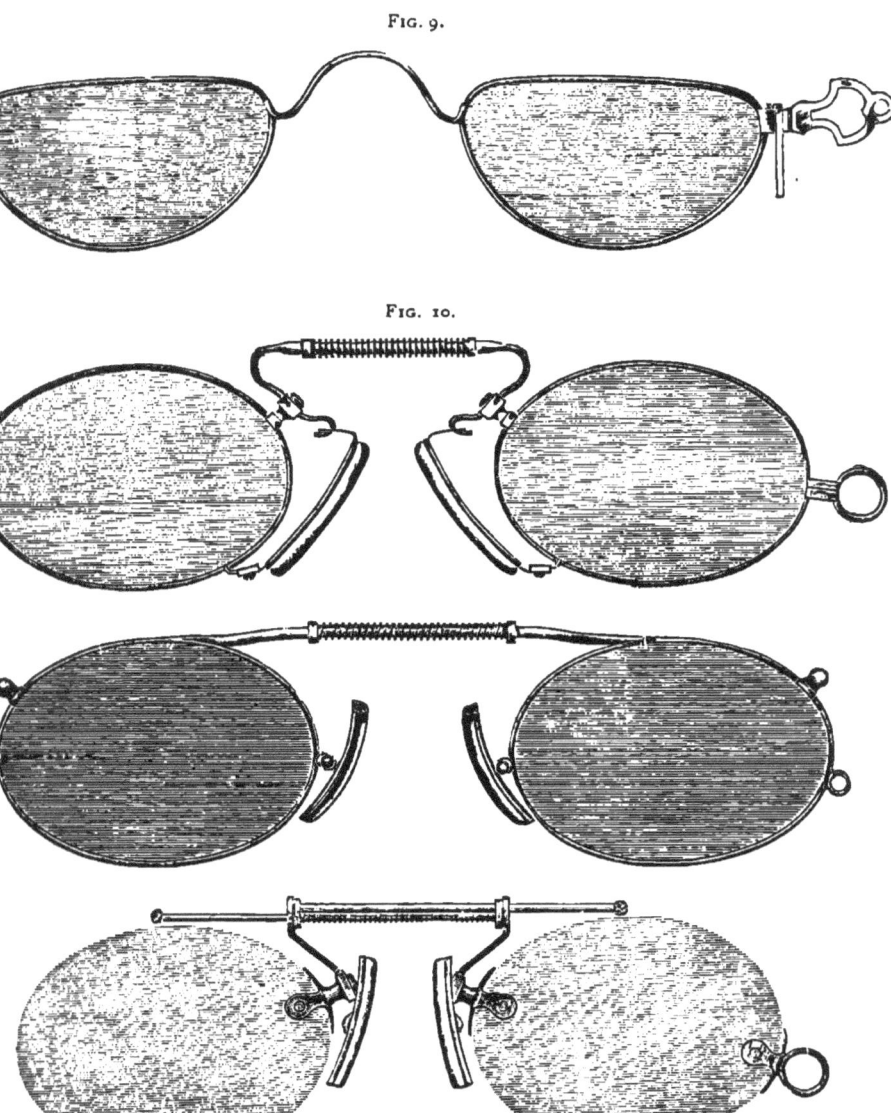

Forms of Rigid Frame or "Bar Spring" Eyeglasses.

was mentioned, the chief objection to the latter is that they allow a displacement of the axis of a cylindrical, or base of a prismatic lens. This is a necessary concomitant of their being joined by a spring instead of a rigid bridge. More or less ingenious frames have been made in which the glasses are rigidly joined, and the spring placed in some other position (Fig. 10); their weight and cumbersome appearance have, however, so far prevented their becoming popular. Another difficulty arises from the fact that it is impossible to always place the glass at the proper distance

FIG. 11.

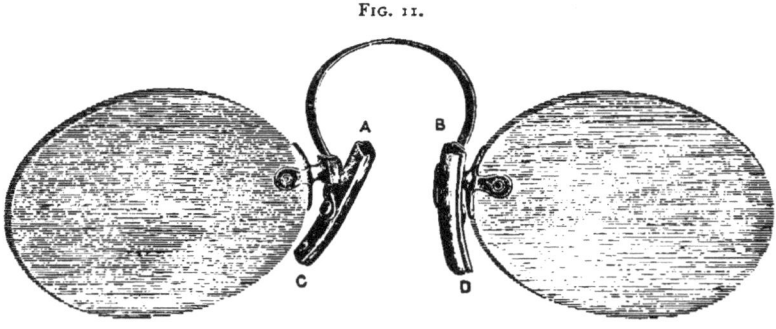

from the eye, since the frame must be placed at that point where it obtains the best grip upon the sides of the nose. This has been in great measure overcome by what is called the "offset guard," in which the nose-pieces are placed back of the plane of the glasses, instead of in the same plane, as formerly. When the glasses contain no prism or cylinder, or only weak cylinders, a well adjusted eyeglass with "offset guards" is fairly satisfactory. When made frameless, as in Fig. 11, it is the most modern, and certainly the most handsome mounting we have to offer our patients.

Spectacles for Cosmetic Effect.—Something may legitimately be done, at times, in the way of improving the appearance of a patient by the application of glasses. The blind whose eyes are not only sightless, but unsightly, very commonly hide them behind colored glasses. Neatly fitting spectacles with large eyes of ground glass render the appearance of such persons less lugubrious. When one eye is useless for vision, and at the same time small, and the orbit undeveloped, a gratifying improvement in the appearance of the patient may be attained by placing before the shrunken eye a convex glass of sufficient strength to magnify it to the size of its fellow. The condition known as epicanthus can generally be removed by wearing eyeglasses whose nose pieces draw just enough on the inner canthi to smooth out the offending fold of skin. As the subjects of epicanthus are generally flat-nosed, it may be necessary to furnish the eyeglasses with a pair of hook temples to keep them in place. Since operations for this disfigurement are so unsatisfactory, such an appliance is probably the best treatment we can advise in case the trouble is not outgrown.

II. THE PRINCIPLES OF SPECTACLE FITTING.

We have now to consider the essential principles of placing glasses before the eyes. The usefulness of spectacles depends almost as much upon the fidelity with which these principles are carried out as it does upon a careful correction of the errors of refraction.

Centering and Decentering.—By the visual axis, or, in English, the line of sight, is meant a line from the yellow spot of the retina through the nodal point of the eye to the object sighted.

By the principal axis of a lens we mean a line passing through the optical center of the lens (the thickest part, if the lens is convex; the thinnest if concave) at right angles to its surfaces.

The geometrical center of a spectacle glass may be shortly said to be that point on its surface which is equally distant from the extremities of the figure to which it is cut. The principal axis of the lens may or may not pass through this latter center.

We habitually regard as the normal position for glasses one in which, when the eyes are looking at a distant object, the visual axes correspond exactly in position with the principal axes of the lenses, and together they pass through the geometrical centers of the spectacles. In other words, the geometrical center of the spectacle eye and the optical center of the spectacle lens coincide, and the center of the pupil for each eye lies directly behind them. Regarding decentering, some confusion is apt to arise because the word is used in two different connections. If the visual axis pass to the temporal side of the optical center of a glass held before an eye, then, with respect to that eye, the glass is said to be "decentered in." If the visual axis pass to the nasal side of the optical center of the glass, the latter is "decentered out." Similarly a glass may be decentered in any other direction. When speaking of spectacles, however, without reference to the eyes of the wearer, they are said to be "decentered in," when their optical centers lie to the inner side of their geometrical centers; "decentered out" when the optical centers

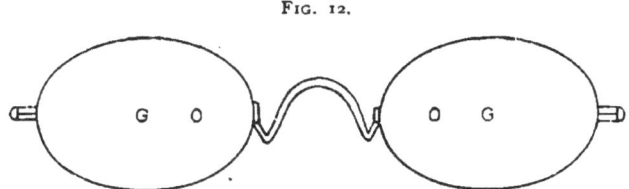

FIG. 12.

SPECTACLES WITH LENSES DECENTERED IN.

G G show the position of the geometrical centers; O O that of the optical centers.

are to the external side of the geometrical centers, etc. A glance at Fig. 12, which represents a pair of spectacles decentered in, will make clear what is meant.

From the above it will readily be seen that when it is desired that a patient wear decentered lenses, the effect may be obtained in either of the two ways: first, by decentering the lenses in their frame; second, by displacing them, together with their frames, from what I have described as the normal position. The first method has the disadvantage of increasing the weight of the glass, while the second limits the field of binocular vision. In practice, the second method should be employed to the greatest extent possible without unduly interfering with binocular vision for the distance at which the spectacles will be used, and, should still farther decentering be required, the method first mentioned should be brought into service. For instance, suppose we wish to order glasses with each lens decentered in 8 mm. This would mean that the optical centers are to be 16 mm. nearer together than the patient's pupils. Let us suppose that by a careful consideration of the distance for which the glasses are prescribed, of the distance at which they must be placed in front of the eyes, and of the size of the spectacle eye used, we find that the frame can only be made 10 mm. narrower than normal without the outer rims of the "eyes" becoming annoying. This leaves 6 mm. to be obtained by decentering the glasses in their eye-wires. If the distance between the patient's pupils were 60 mm., we would order the distance between the geometrical centers of the spectacle eyes to be 50 mm., and each eye to be decentered in 3 mm.

Prismatic Effect of Decentering.—It is to obtain a prismatic effect from spherical lenses that decentering is generally ordered, since a decentered lens is identical with a lens of the same strength combined with a prism. This is graphically shown by Figs. 13 and 14, the latter of which represents a section of a decentered lens, which will readily be seen to be precisely the same as the result would be if the normally centered lens shown in Fig. 13 were split into halves and the prism $b\,a\,c$ introduced between them.

The size of the glass disks from which spectacle lenses are ground will not allow of more than about 2 mm. of lateral decentering for a No. 1 "eye;" 3 mm. for Nos. 2

THE PRINCIPLES OF SPECTACLE FITTING. 47

and 3; and 4 mm. for No. 4. Vertically, they may be decentered much more. When ordered to decenter laterally more than this, or to furnish a prismatic effect greater than can be obtained by this much decentering, the optician first manufactures a prism of the requisite strength, and then grinds spherical surfaces upon its two faces. It is, therefore, of not much importance whether, in ordering a sphero-prismatic combination, we express the prismatic element in degrees of the refracting angle, or in millimeters of decentration of the lens: the optician produces the glass by whichever method is most convenient.

SHOWING THE PRISMATIC EFFECT OF DECENTERING.

The optical center, O, in Fig. 13, coincides with the geometrical center, G, in Fig. 14, which represents a decentered lens of the same spherical curvature; O has been removed towards the base of the virtual prism *bac*. (*After Maddox.*)

The stronger the lens, the less decentering it requires to produce a given prismatic effect, and where the combination desired is that of a strong lens with a weak prism, the more accurate practice probably is to order the lens decentered the requisite number of millimeters. For this purpose a table of equivalents, such as is given below, is necessary. To use it, we find in the first column the strength of the lens used, and on a level with this, in the column at whose head stands the strength of the prism required, is given in millimeters the amount of decentration necessary.

TABLE II*—DECENTERING EQUIVALENT TO A GIVEN REFRACTING ANGLE (INDEX OF REFRACTION, 1.54.)

Lens	1°	2°	3°	4°	5°	6°	8°	10°
1 D,	9 4	18.8	28.3	37.7	47.2	56 5	75.8	95 2
2	4.7	9.4	14 1	18.8	23.6	28 2	37.9	47.6
3	3.1	6.3	9.4	12.6	15.7	18.8	25.3	31.7
4	2.3	4.7	7.1	9.4	11.8	14.1	18.9	23.8
5	1.9	3.8	5.7	7.5	9.4	11.3	15.2	19.
6	1.6	3.1	4.7	6.3	7.9	9.4	12.6	15.9
7	1.3	2.7	4.	5.4	6.7	8.1	10.8	13.5
8	1.2	2.3	3.5	4.7	5.9	7.1	9.5	11.9
9	1.	2.1	3.1	4.2	5.2	6.3	8.4	10.5
10	.9	1.9	2.8	3.8	4.7	5.6	7.6	9.5
11	.9	1.7	2.6	3 5	4.3	5 1	6.9	8.7
12	.8	1.6	2.4	3.1	3.9	4.7	6.3	7.9
13	.7	1.4	2.2	2.9	3.6	4.3	5.8	7.3
14	.7	1.3	2.	2.7	3.4	4.	5.4	6.8
15	.6	1.3	1 9	2 5	3.1	3.8	5.1	6.3
16	.6	1 2	1 8	2.4	3.	3.5	4.7	6.
17	.6	1.1	1 7	2.2	2.8	3.4	4.5	5.6
18	.5	1.	1.6	2.1	2.6	3.1	4.2	5.3
19	.5	1.	1.5	2.	2.5	3.	4.	5.
20	.5	.9	1.4	1.9	2.4	2.8	3.8	4.8

A cylindrical lens, or the cylindrical element of a spherocylindrical lens, when decentered in a direction vertical to its axis, acts as a spherical lens of the same strength. Thus, a +2. Sph. ○ +1. Cyl. axis vertical, decentered horizontally, would have the same prismatic effect as a +3. Sph. treated in the same way. As the axis is inclined toward the direction of decentration, the prismatic effect of the cylinder diminishes, and disappears when they coincide. Thus, a +2. Sph. ○ +1. Cyl. axis horizontal, decentered horizontally, would have merely the prismatic effect of a +2. Sph. so treated.

---

* Jackson: "Transactions of the American Ophthalmological Society," 1889.

THE PRINCIPLES OF SPECTACLE FITTING.

Normal Lateral Centering. — In proportion as the prismatic effect of decentered lenses is a valuable property where this effect is desired, it has to be guarded against in those cases which do not require it, to which number belong, of course, the great majority of the cases we are called upon to treat. If the objects looked at through spectacles were always situated in the same direction and at the same distance, fixing the position proper for the centers would be a simple matter; but, in the movements of the eyes, each pupil roves over a territory some 18 mm. (¾ in.) long by 15 mm. broad. When the eyes are directed toward a distant object the centers of the pupils are about 60 mm. apart, and on convergence only 56 mm., so that the proper adjustment of spectacles is a series of compromises between that proper for the position of the eyes in which the glasses will be most used and other positions in which they will be less used. Of course, the position in which they will be most used must receive the greatest consideration.

The proper position for the centers of " distance " glasses has already been stated. When glasses are to be used for near work only, they should be decentered " in " two or three millimeters on each side from this " normal " position, as such glasses being never used in that position, but only when the visual axes are converged, would otherwise never be rightly centered. What amounts to the same thing, and is more often done, is to make the front of the near spectacles four or six millimeters narrower than if they were intended for distant vision : four millimeters narrower for a working point of 15 inches; six millimeters narrower for one of 10 inches. Concerning the centering of glasses which are worn constantly, no rule for all cases can be laid down, since accurately centering for any one distance is decentering for every other. Fortunately, as a glance at Table II will show, it is only with lenses of high power that a considerable amount of prismatic effect is developed by slight decentering. Where such glasses must be worn constantly by a person who spends several hours daily at near work, they should certainly be slightly decentered inward.

The distance between the geometrical centers is regulated by the size of the spectacle eyes and the width of the space between them occupied by the bridge. Where the interpupillary distance is short, as in children, opticians are apt to make the eyes of the spectacles so small as to interfere seriously with the field of vision through them. With the saddle bridge there is no difficulty in diminishing the space between the spectacle eyes without interfering with the form of that part of the bridge which is applied to the nose, and the required adjustment should be made in this way; leaving the spectacle eyes of good size.

Normal Vertical Centering.—The glasses require, further, to be so placed that the points where the wearer's visual axes penetrate them shall neither be above nor below the centers. This adjustment is readily seen to depend upon the relative height of the bridge of the spectacles and the bridge of the nose at the point where the spectacles rest. The higher the spectacle bridge, the lower will the glasses stand upon the patient's face, and vice versa.

On the bridge of nearly every nose there may be felt a point at which the narrow, upper portion of the nasal bones gives place rather suddenly to the broader lower portion. Just here, in what has been called the "natural" position (A, Fig. 15), the bridge of the spectacles tends to rest, and the attempt to make it remain at any other point will not be very successful. In distance spectacles, then, the bridge

THE PRINCIPLES OF SPECTACLE FITTING.

should be made of such height that when resting at this natural position the centers of the spectacle eyes are at the same height as the centers of the pupils when the patient looks straight forward. When the glasses are to be used for near work only, their bridge should be made about 2 mm., or ⅛ inch higher than otherwise, allowing the centers to drop that much lower, as the wearer's eyes will nearly always be directed to objects below their own level.

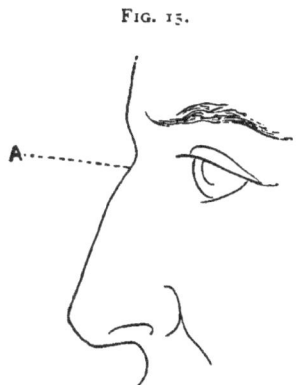

FIG. 15.

Distance of the Glasses from the Eyes.—As a rule, the glasses should be placed just far enough from the eyes to escape the lashes in the act of winking. If the lashes touch the glass the latter quickly becomes soiled, and to the spectacles is, moreover, attributed any falling out of the lashes which may occur. Some persons, however, with myopia of high degree, prefer the glasses to be placed as close to the eyes as possible, regardless of the lashes, because of the larger clear images which they thus obtain. This adjustment of the glasses depends upon the relation of the top of the spectacle bridge to the plane of the glasses.

Where the eyes are deep set, or the nose of the aquiline type, the top of the spectacle bridge must be in front of the plane of the glasses, or, as it is shortly called, "out" (Fig. 16). When the bridge of the nose is low and the eyes relatively prominent, as in the negro, Chinese and children, the top of the bridge must be back of the plane of the glasses, or "in," as represented in Fig. 17.

Perpendicularity of the Plane of the Lenses to the Visual Axis.—A very important requirement, and one not sufficiently regarded in the fitting of frames, is that the plane of the correcting lens when in use shall be as nearly

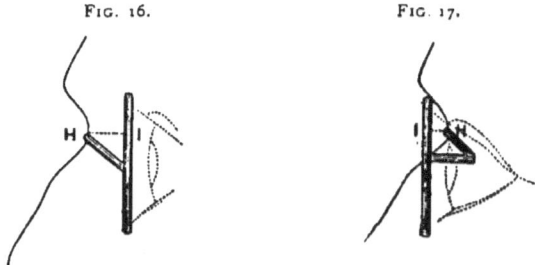

FIG. 16.   FIG. 17.

as possible perpendicular to the visual axis. The stronger the lens the more important is this detail, whose warrant lies in the fact that the refractive value of a given lens placed obliquely to the visual axis is no longer that indicated by its number, but is that of some other, stronger lens. A cylindrical lens so placed acts simply as a stronger cylindrical lens; a spherical lens, however, as a stronger spherical lens combined with a cylindrical lens with its axis at right angles to that about which the lens is rotated.

The results of the investigations of himself and others, of the effect of the obliquity of a lens to an incident pencil of rays, was summarized by Dr. Edward Jackson, in a paper read before the American Medical Association in 1877, and their practical application to this part of our subject pointed out. From that communication the following table is extracted. It gives in the first column the degrees of obliquity at intervals of 5° up to 45°. In the second column is shown the refractive value of a 1. D. cylindrical, in the third that of a 1. D. spherical lens so inclined.

TABLE III.

Obliquity of the Lens.	Refractive power of a 1 D. Cylindrical Lens so Placed.	Sphero-Cylindrical Equivalent of a 1 D. Spherical Lens so placed.
0°	1. D. cyl.	1. D. spherical.
5°	1.01 "	1.00 sph. ⌒ 0.01 cyl.
10°	1.04 "	1.01 sph. ⌒ 0 03 cyl.
15°	1.10 "	1.02 sph. ⌒ 0.08 cyl.
20°	1.17 "	1.04 sph ⌒ 0.13 cyl.
25°	1.30 "	1.06 sph. ⌒ 0.24 cyl.
30°	1.44 "	1.09 sph. ⌒ 0.36 cyl.
35°	1.69 "	1.12 sph. ⌒ 0.56 cyl.
40°	2.01 "	1.16 sph. ⌒ 0.83 cyl.
45°	2.46 "	1.22 sph. ⌒ 1.24 cyl.

To fulfil this requirement of perpendicularity to the visual axis, the lenses of spectacles used only for distance should lie in a vertical plane; that is, they should face directly forward, as shown in Fig. 18. Since the visual axes are directed downward and forward when near work is done below the level of the eyes, glasses for near must face downward and forward, as in Fig. 19, in order that the plane in which they lie shall be perpendicular to those axes. Furthermore, in viewing near objects the visual axes are directed inward and toward each other. This will require the glasses to face inward also, as represented in Fig. 20,

Fig. 18.

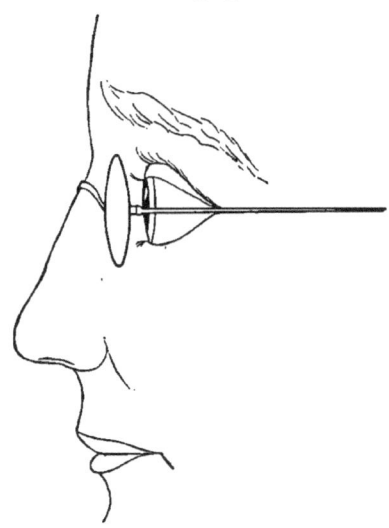

Fig. 19.

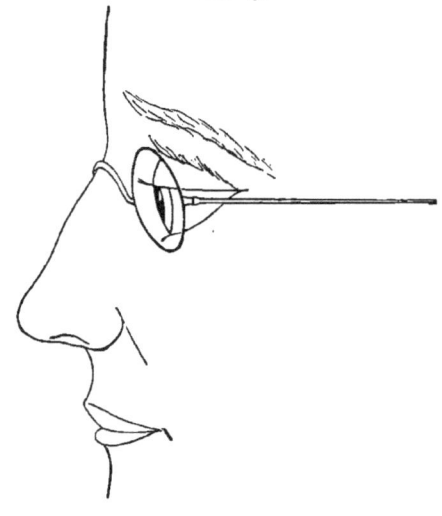

so that they come to lie in different planes, instead of in the same plane as formerly.

When "constant" glasses are prescribed, the lenses should be placed midway between the proper facing for near and that for distance glasses. Then, though the lens is not exactly properly inclined either for distant vision or near work, the result of such slight obliquity to the visual axis is unimportant, since, as a reference to Table III will show, it is only in the higher degrees of obliquity that the increase in power, and especially the development of cylindrical effect from spherical lenses is rapid. Moreover, by slightly bending the neck a moderate

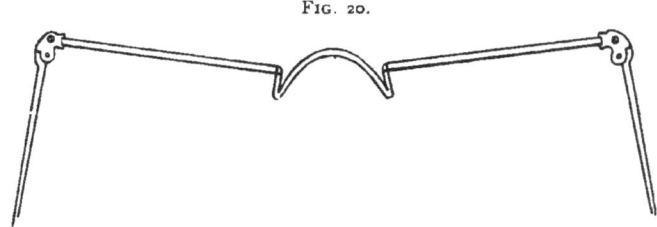

FIG. 20.

degree of obliquity of the glasses to the visual axes may be removed without discomfort to the wearer.

The position of bifocal glasses should also be between that proper for near and for distance glasses, but nearer that of the stronger glass. This will generally be the near glass, as convex bifocals are much more frequently prescribed than concaves, and such glasses should face only a little less downward than glasses intended entirely for near work. When concave bifocals are worn, however, they should face more forward and much less downward.

The angle which the plane of the glasses makes with the plane of the wearer's face depends entirely upon the angle formed by the plane of the glasses and the temples of their containing frames. Thus, when the temples are perpendicular to the plane of the glasses, as in Fig. 18, the latter will face forward and not at all downward. They may be made to face downward to any required degree by simply turning down the temples at the points where they are hinged to the end pieces. These must be equally turned down, however, as where only one is turned down, or one more so than its fellow, the result is not to make the glasses face downward, but to make the glass on the side of the lower temple ride higher on the face than its fellow.

Periscopic Glasses.—In the effort to further apply the law requiring that the plane of the lenses shall be perpendicular to the visual axes, we are met with the fact that with biconvex and biconcave lenses this relation is only strictly possible within a comparatively limited area surrounding the optical center of the lens. When the wearer looks through the periphery of his glasses the visual axes will pierce the lenses obliquely, and the refractive value of the latter will, of course, be governed by all the laws of tilted lenses. For instance, when the wearer of an ordinary convex lens looks through it near the edge, the optical effect of the glass before his eye is that of a stronger convex lens combined with a cylindrical lens; the axis of the latter depending on the part of the periphery pierced by the line of sight. In weak lenses, the slight inaccuracy of vision produced in this way is of small moment, but where the strength of the lens used is greater than about 2. D. the patient's field of accurate vision is greatly reduced in size, and in viewing objects not directly in front of him, he is obliged to perform wide motions of the head in order to be able to see them through the central portion of his glasses. This is especially true of cases of aphakia, where, of course, very strong lenses are generally necessary. To escape or lessen these disadvantages, strong spherical lenses should be, and generally are, made in the form of a meniscus, which when placed with its convex surface *from* the eye constitutes a periscopic glass. The ideal of this form of lens may be defined as a glass in which the center of curvature of one surface coincides with the center of rotation of the eye, and that of the other surface approaches it as closely as the required strength of the glass will permit. In such a glass the visual axis will always be perpendicular to the first surface, and nearly so to the second, at whatever point it pierces the glass, and in whatever direction the eye may be turned.

When a cylindrical or sphero-cylindrical lens is required, the best form of glass is the toric lens described on page 33. These lenses have, however, never been manufactured extensively, and the process of their manufacture, as well as the lens itself, being patented in this country, their cost is considerable. By transposing the usual formula, however, there may be obtained from any optician a sphero-cylindrical lens which approaches the periscopic form, and is certainly superior to one ground after the usual method. For illustration, if one desires to order $+$ 2. D. Sph. $\bigcirc$ $+$ .75 D. Cyl. Ax. 90°, the formula may be transposed and the order written for $+$ 2.75 D. Sph. $\bigcirc$ $-$.75 D. Cyl. Ax. 180°. This glass, though optically of the same strength as the first, would have an approach to the periscopic form if placed with the cylindrical surface next the eye. The field of accurate vision would gain in all directions, especially in the vertical one, in which diameter, however, its enlargement is not of so much consequence as it is laterally. Aphakic eyes offer the best field of usefulness for this practice, as in them we have generally to deal with a high hyperopia, and often with hyperopic astigmatism requiring for its correction a convex cylinder with its axis horizontal. Let us suppose that after a cataract extraction we wished to order + 10. D. Sph. ◯ + 6. D. Cyl. Ax. 180°. With this lens, accurate vision would be limited to a vertical oval field situated directly in front of the patient, beyond the confines of which all objects would appear distorted by various cylindrical effects. We would, therefore, transpose the formula into + 16. D. Sph. ◯ — 6. D. Cyl. Ax. 90°, and this glass will be likely to give the patient much more satisfaction than the other would have done, as with it he obtains a very good lateral field.

III. PRESCRIPTION OF FRAMES.

In order to prescribe the frames for a pair of spectacles, we must, after measuring the face or a frame which fits, record the dimensions of the frame we desire to order. The essential measurements are the intercentral distance, or width of front, and the three dimensions of the bridge. This list may be extended to include the measurement of the angle formed by the bridge and the plane of the lenses, that formed by the temples and the plane of the lenses, the distance between the temples an inch back of the glasses, and the distance from the hinge of the temples to the top of the wearer's ear. All these details are, however, so ready of adjustment, and the trouble and uncertainty of their prescription are so great, that in my judgment they are better left until the frame is received from the maker and we are ready to adapt it to the patient's face. The distance between the centers of the spectacle eyes is best obtained by measuring upon the face the distance between the centers of the pupils; the other dimensions of the frame, however, are more easily obtained by trying on a sample frame and taking the measurements from this, estimating any change which may be necessary. To do this requires about a half-dozen sample frames whose bridges are of different dimensions; also a rule graduated in millimeters, or sixteenths of an inch. I have had made for this purpose a rule which I think facilitates the work. As represented in Fig. 21, it has upon one side three scales graduated in millimeters and conveniently placed for taking the

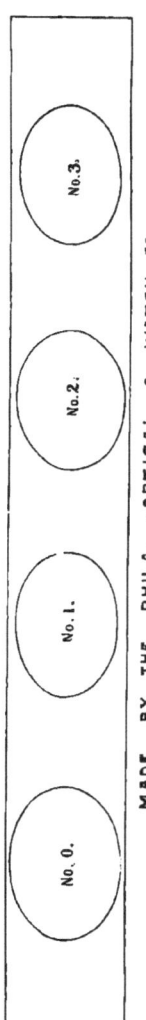

Fig. 21.

Made by the Phila. Optical & Watch Co.
Obverse and Reverse of a Convenient Rule for Measuring Spectacle Frames; One-half Actual Size.

PRESCRIPTION OF FRAMES.

different dimensions of the frame, while on the reverse side are several ovals showing the principal sizes of spectacle eyes. Some of the uses of these scales are shown in Figs. 22, 23, 28, and 29; to avoid confusion, one scale only is drawn in each diagram.

Philadelphia, 189

Name of Patient,

℞.

O. D.

O. S.

Unless otherwise specified, furnish the following: Medium length temples; saddle bridge; No. 2 eyes. Dimensions are given to middle of wires. Dimensions given are in millimeters.

Frames of $\begin{array}{c}gold\\steel\end{array}$ *Catalogue No.* ....

Interpupillary Distance ....

Bridge $\begin{cases} Height .... \\ Width\ of\ Base .... \end{cases}$ *Top* $\begin{array}{c}in\\out\end{array}$

.... *M. D.*

A prescription blank such as that here given indicates what measurements are required, and will be found useful in practice. The upper part is for the lenses, the lower part for the frames.

To Obtain the Interpupillary Distance, with which the first dimension of the frame, the distance between the geometrical centers (A to B, Fig. 22), is generally identical, the physician seats himself facing the patient in a good light, the latter being directed to look straight before him

FIG. 22.

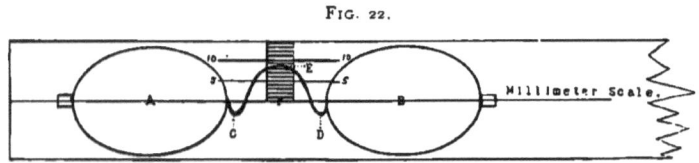

at some distant object. The measuring rule is placed before the patient's eyes, as close to them and as far from the physician's eyes as possible. The zero of the scale being placed opposite the center of one pupil, the center of the

FIG 23.

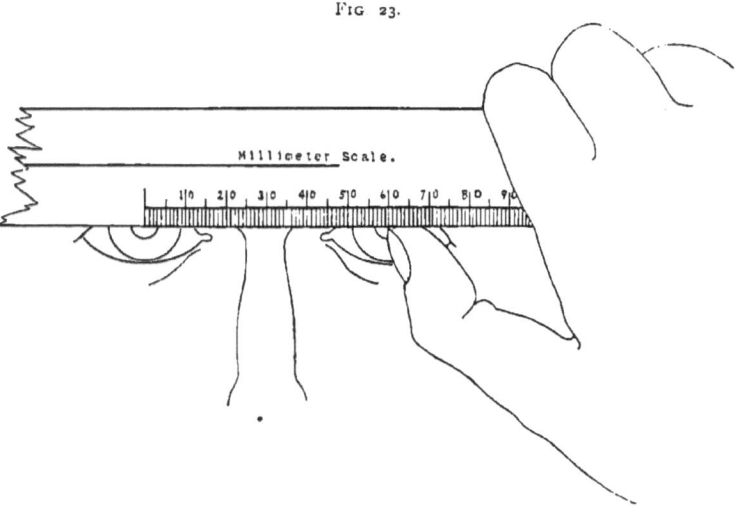

other may be marked by the physician's thumb nail, as represented in Fig. 23, and the distance between them read off the scale. This distance seldom varies more than 5 mm.

PRESCRIPTION OF FRAMES. 63

from 60 mm., or 2⅜ in. It will be observed that as the physician's eyes are less than the length of his arm away from the patient's face when this measurement is taken, in fact about two feet away, the marks upon the rule, though apparently opposite the pupils, will in reality be a little within the centers; so that the distance obtained will be a little less than it should be. When the physician's eyes are two feet away from those of the patient, and the rule is one inch away from them, the error in measuring an interpupillary distance of 60 mm. by this method is almost

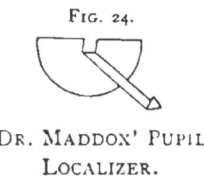

Dr. Maddox' Pupil Localizer.

Fig. 25.

The Pupil Localizer in Use.

exactly 2 mm. This amount should, therefore, be added to the apparent interpupillary distance to obtain the true one.

The measurement obtained in this way is sufficiently accurate for most purposes, but if a greater degree of accuracy be desired in any case it may be attained by means of the little device suggested by Dr. Maddox, which is represented in Fig. 24. This is to be placed before one of the patient's eyes in an ordinary trial frame having a graduated bar for showing the distance of each geometrical center from the middle of the bridge. The gaze of the observed and that of the observing eye being directed to each other's pupils, the two sights of the implement are brought into line between them, as shown in Fig. 25. The same procedure is then gone through with for the other eye, and the distance of the second pupil from the median line of the face, as regis-

Fig. 26.

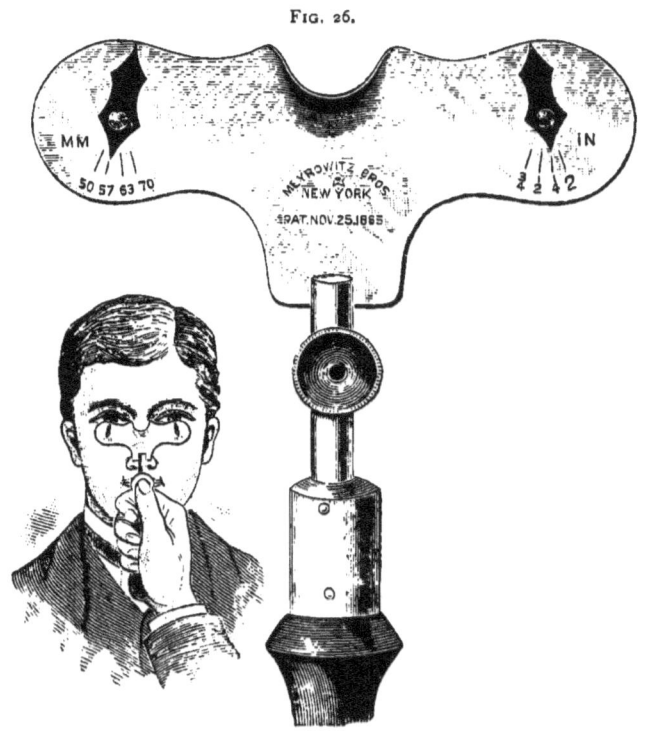

tered by the trial frame, is added to that of the first, to obtain the interpupillary distance. This procedure is also of advantage in revealing and measuring any difference in the distance of the pupils from the median line, due to asymmetry of the face. The use of a trial frame for making accurate

PRESCRIPTION OF FRAMES. 65

measurements requires the bestowal of considerable attention to see that the support of the nose piece is vertical, the joints close and tight, and the markings correct; otherwise it may readily introduce the errors its use is intended to obviate. There are, in the shops, many special forms of the "pupilometer" constructed on the principle of a rule held before the eyes and a single sight for each pupil. Two of these are shown in Figs. 26 and 27. The interpupillary distance as registered by them requires, of course, the same correction as does that obtained by the simple graduated rule.

Height of the Bridge.—This is the distance of the top of the bridge above a line joining the centers of the

FIG. 27.

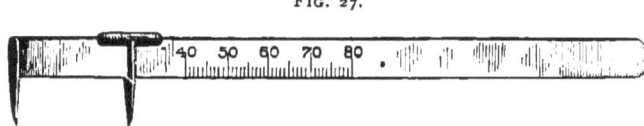

lenses. In Fig. 22, it is the distance from E to F, which is the height of E above a line joining A and B; not the height of E above a line joining C and D, which is sometimes erroneously supposed to represent the height of the bridge.

If a rule be held horizontally before the patient's eyes, with the lower edge touching the nose at the natural position for the spectacle bridge, the height of this edge of the rule above the pupil on either side will show at a glance about how high the top of the future bridge must be. We may then select from our sample frames that one whose bridge corresponds most nearly with this supposed height, and being sure to place it in the natural position, we carefully note whether the pupils are above or below the centers of the eyes of the frame. If they are below these centers, sufficient must be added to the height of the bridge now upon the face to allow them to coincide; if the pupils are above the centers, a corresponding subtraction from the height of the trial bridge must be made. Each sample frame may have its dimensions attached to it, or any frame may be used as a fitting frame and afterward measured. To measure the height of a bridge the glasses are laid upon a sheet of ruled paper, or other object offering a convenient straight line, in such a

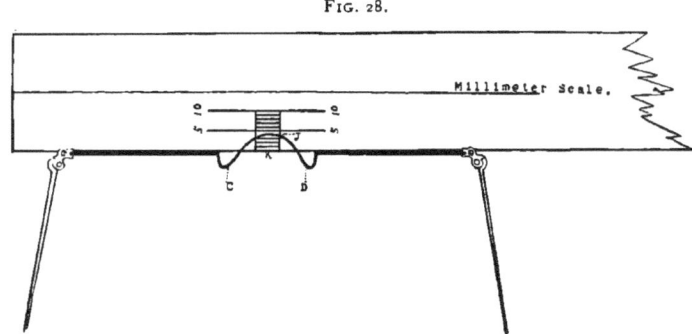

Fig. 28.

way that the line passes through the geometrical centers of the eyes, or, what is the same thing, through the joints of the end pieces on each side (Fig. 22). The height to which the bridge projects above this line is then readily measured. It is seldom greater than 10 mm., and in rare cases may be a minus quantity, the top of the bridge being below the level of the centers of the lenses.

Relation of the Top of the Bridge to the Plane of the Lenses.—The measurement required to express this relation is that from J to K in Figs 28 and 29; not the distance of J in front of a line joining C and D, as might

PRESCRIPTION OF FRAMES.

be supposed. This measurement is also shown at H I, Figs. 16 and 17; it is obtained by a procedure similar to that just described for obtaining the height of the bridge. The rule being placed across the nose at the natural point, and the patient requested to wink, it may readily be seen whether the lashes touch the edge of the rule. If they do, the top of the bridge of the future spectacles must be back of the plane of the glasses, or "in." If they do not, we note how much nearer, if any, the edge of the rule might be brought without their touching, and so obtain a guide to the distance the top of the bridge should be in

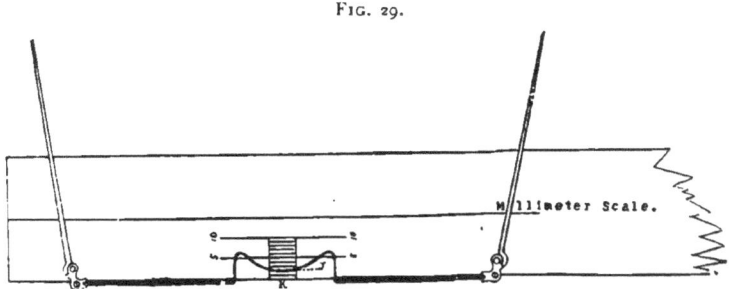

FIG. 29.

front of the plane of the lenses, or "out." The fitting frame which comes nearest to the requirements of the case in this particular is then placed upon the face, when by viewing it from above or from the side it can quickly be seen just how much change, if any, is needed to place the glasses a little beyond the reach of the lashes. The method of measuring the distance of a bridge in or out is so plainly shown in Figs. 28 and 29 that special explanation is unnecessary. They seldom measure more than 4 mm. out or 3 mm. in.

Width of Base.—The measurement from C to D, Fig. 28, is obtained, like the others, by measuring a bridge which fits or estimating the change necessary in one which does not. This dimension is usually from 16 mm. to 20 mm.

This method of obtaining the dimensions of the bridge required may seem tedious and uncertain in the description; in use it is not so, and after trial I think will be found preferable to any special device so far invented for recording the measurements. These, after shifting of screws and bending of wires, leave one to estimate what changes are required just as might have been done without their aid. Moreover, the heavy parts and lost space in joints of trial frames may readily conceal an error of 2 mm., or even 3 mm. in some measurements; the large, round eyes with heavy rims will not go under the brows, so that the in-out measurement of the bridge must frequently be guessed at; and the relation of the upper part of the eye wires to the brows is not shown. In fact, they introduce, in my estimation, quite as many sources of error as they eliminate.

Where the face is unsymmetrical no exact rules of procedure can be given, and considerable ingenuity may be required to fit a frame to such a face. If the nose is very peculiar, or one side of its bridge markedly steeper than the other, it may be of advantage to take an outline of the bridge at the natural position by bending a piece of lead wire to fit accurately and marking the outline of this upon the prescription-blank, or sending the wire itself to the spectacle maker. Sometimes the brows are overhanging and the eyes deep set; so that the glasses cannot be properly centered before the eyes and placed close to them without the upper part of the rims burying themselves in the brows. In such cases the glasses should be decentered upward in their frames and the bridge made sufficiently high to bring the optical centers opposite the pupils. Though the patient will then look through the upper part of his glasses, his field of vision will not be any more limited than is already the case because of the overhanging brows.

Prescription of Eyeglasses.—The dimensions which it is usual to furnish in prescribing eyeglass frames are the interpupillary distance, of course, with the distance between the two upper and the two lower ends of the nose pieces when they are in place on the face. (A to B, and C to D, Fig. 11.) These measurements alone will not insure a good fit in the frames, since neither the contour of the sides of the nose to which the guards are applied, the vertical centering of the lenses, nor the distance of the latter from the eyes are taken into account; but the same remark applies here as to the minor dimensions of spectacle frames, namely: that it is more simple, certain, and expeditious for the surgeon to make these adjustments in the frames themselves than to prescribe what the manufacturer shall do for him. Fortunately, eyeglass frames admit of great variation by bending their different parts, and being put together with screws, these parts are quickly interchangeable. Almost the only thing about them which admits of no adjustment is the length of the spring, and, it is well for one who prescribes many eyeglass frames to have a series of such springs at hand from which to replace one which may be found too long or too short.

IV. INSPECTION AND ADJUSTMENT OF SPECTACLES AND EYEGLASSES.

Ordinary prudence demands that the prescriber of glasses make a careful examination of the manner in which his directions have been carried out, since neglect of this precaution may nullify the results of the most painstaking correction of the refraction. If the surgeon himself furnish the spectacles, it is doubly incumbent on him to make a thorough inspection of glass and frame, and to carefully adjust the latter so as to be entirely comfortable to the wearer. Then, too, it is not enough that the frames correctly perform their function at first; they must continue to do so. Should there be no optician in his neighborhood, the surgeon will be called upon to bring to a proper shape frames which have passed through all sorts of accidents, and it is better that he should do this work than entrust it to less competent hands.

Proving the Strength of Lenses.—The focal strength of a convex lens may be directly measured by finding the distance at which it brings the sun's rays to a focus. To do this, the rays which have passed through the lens are simply caught upon a piece of paper, or other screen, the two being held in such relationship that the image of the sun formed on the screen is round. The screen is then to be moved back and forth until the point is found at which this image is smallest, and the distance of such point from the lens is the focal length of the lens. To learn the strength of the lens in diopters, we divide 100 centimeters

INSPECTION AND ADJUSTMENT OF SPECTACLES. 71

(one meter) by the focal length expressed in centimeters, or 40 inches (about one meter) by the focal length expressed in inches. For instance, if we found the focus of the lens under examination to be distant 10 in., or 25 cm., from the lens, 40 in. divided by 10 in., or 100 cm. divided by 25 cm., will alike give a quotient of 4, and the lens measured was, therefore, a + 4. D.

The focal length of a concave lens may be similarly measured by combining it with a stronger convex lens and then measuring the strength of the resulting weaker convex. The strength of the original convex used being known, we have only to subtract from it the weak convex resultant to find the strength of the concave with which we are dealing. The focal length of convex and concave cylindrical lenses may be measured in the same way as the corresponding sphericals, it being only necessary to observe that the parallel rays of light after passing through a convex cylindrical lens are arranged in the form of a line at the focus of such lens; not brought to a point, as is the case with convex sphericals.

Phacometers.—Such methods as the one described above are, however, too tedious for ordinary use, though quite elaborate contrivances called phacometers have been devised on this principle. A lens measure constructed on an entirely different idea has lately appeared, the invention of Mr. J. T. Brayton, of Chicago. Fig. 30 shows the size and appearance of the instrument, as well as the method of its use. Of the three steel pins which project from its top the two outer ones are fixed, while the central one moves up and down easily but is held up by a spring. On pressing the surface of a spherical lens squarely against these points, the central one will be depressed until they all three touch the glass; the curvature of the surface of the lens determining the amount of such depression. The motion being transferred through a rather simple mechanism to the hand upon the dial, this travels over a scale which shows in diopters and in inches the strength of the lens corresponding to the surface tested. The other surface is then to be explored in the same way. If the lens is biconvex or biconcave, the results of measuring each surface

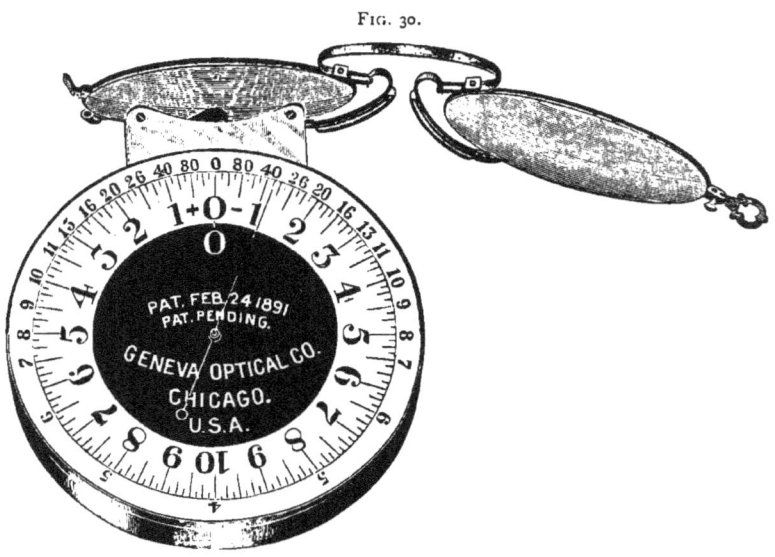

Fig. 30.

separately are added together; if periscopic, the less is deducted from the greater. When used upon a cylindrical surface the hand will stand at zero when the three points are in line with the axis of the cylinder. When the points are placed at right angles to the axis the strength of the cylinder is shown.

Since this instrument indicates the refractive value of a lens from the curvature of its surfaces only, leaving out of account the index of refraction of the material, it is evident that it can be accurate for only one variety of glass. As found in the shops it is adjusted for crown glass, and for lenses of this material it is quite accurate; while its convenience and low price as compared with other phacometers recommend it to favor.

Neutralization of Spherical Lenses.—The method of determining the strength of spectacles which is of most general utility is the well-known one of neutralization. If a convex spherical lens be held about a foot from the eye, and any object, say that part of a window frame where a vertical and horizontal line cross, be viewed through it, any motion given the lens will result in an apparent motion in the opposite direction of the object sighted. That is; if the lens is moved to the right, the object appears to move to the left; if the lens is raised, the object appears to sink. If the same maneuver be employed with a concave spherical glass, the object again appears to move, but this time in the same direction as the motion imparted to the lens. If the lens is moved to the right, the object appears to move to the right also. Here we have the readiest possible means of distinguishing between a convex and a concave lens. Moreover, one gets in this way an idea of the strength of a lens, as the stronger the lens the more rapid is the apparent motion of the object seen through it.

If, continuing the experiment, the two lenses be placed together, with their curved surfaces in apposition, and a trial be made of the effect of moving them before an object, as was done previously with each lens singly, the object will appear: 1, (if the concave lens is the stronger) to move in the same direction as the motion of the glass, but more slowly than before; 2, (if the convex lens is the stronger) to move in the opposite direction to the motion of the glass, but more slowly than before; 3, (if the lenses are of equal strength) to have no motion. Therefore, to find the strength of a spherical lens it is only necessary to combine it in this way with successive lenses of known strength and of the opposite sign until that one is found which neutralizes the apparent motion of objects seen through it. This lens is the measure of the strength of the one tested. This method is accurate within an eighth diopter, or less, for plano-convex and plano-concave lenses; with bi-convex and bi-concave glasses it is only possible to neutralize the

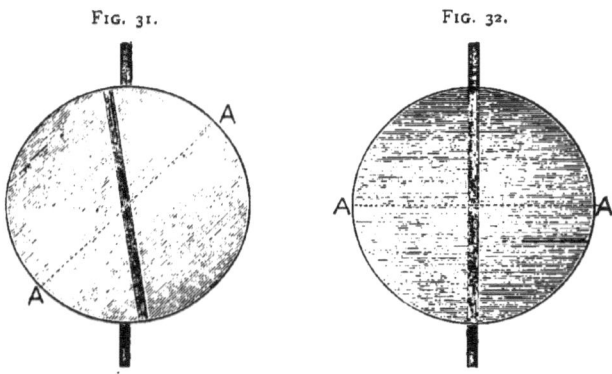

FIG. 31. FIG. 32.

apparent motion near the center of the lens; toward the edges motion will still be visible when the lenses are strong.

Cylindrical lenses may be recognized by viewing through them some object presenting a straight line, say the vertical line of a window sash. If the cylindrical lens be rotated about the visual axis, the portion of the vertical line seen through the glass will appear to be oblique, as compared with that seen above and below the glass (Fig. 31). This oblique displacement takes place in a direction contrary to the rotary motion given the lens if the latter is convex, and in the same direction as the motion if the lens is concave. To ascertain the position of the axis of a cylindrical lens it should be rotated slowly in this manner until the line seen through it appears continuous with that above and below (Fig. 32). This line will then lie either in the axis or at right angles to it. To ascertain which of the latter is the case the effect of motion from side to side is to be tried. If the axis of the cylinder corresponds with the vertical line looked at, motion from side to side produces apparent motion of the object; if, however, the axis lies at right angles to the vertical line no such motion results. In other words, in the direction of its axis a cylindrical lens acts as a piece of plain glass; across its axis it acts as a spherical lens of the same strength. The direction of the axis of a cylindrical lens having been ascertained, its strength may be determined by neutralizing it with a cylinder of the opposite sign, as was explained when speaking of spherical lenses. Care must be taken that the two lenses are so placed that their axes coincide.

A Sphero-Cylindrical Lens is equal in refractive effect to two cylindrical lenses with their axes perpendicular to each other. Having found that axis across which motion is least rapid, we may neutralize the motion with a spherical lens, and holding these two together, proceed to neutralize the motion across the other axis just as if dealing with a simple cylinder. When our object is not to determine the strength of an unknown lens, but to see if the lenses of a pair of spectacles agree with the prescription previously written, we may, of course, shorten the above procedures by picking out from the test case the glass, or glasses, which will neutralize the spectacles if the latter are of the proper strength, and observing whether the apparent motion of objects ceases when they are held together.

Locating the Optical Center. —Every glass before being worn should be examined with regard to the position of the optical center of each lens and the distance of these from each other, as inaccuracy in this important particular is not uncommon. Indeed, in the cheap spectacles which some persons unfortunately buy, proper centering is the exception. In grinding large numbers of lenses by machinery a certain number in each batch are, I believe, always found to be badly centered. These are not returned

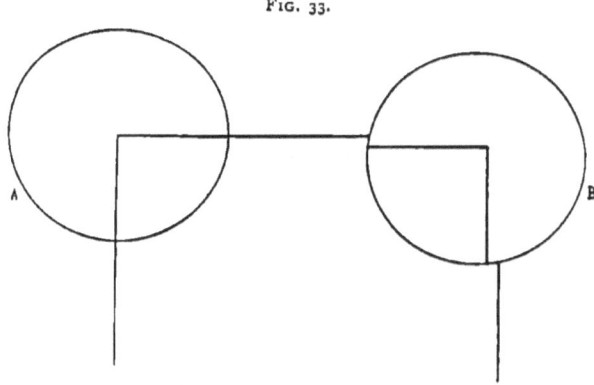

FIG. 33.

to the wheel or the furnace by the thrifty manufacturer, but are graded as second class, or if very bad indeed as third class, and with those which will not pass inspection in other particulars go to make up the trash sold by peddlers.

A simple way to find the location of the optical center is to hold the lens about a foot above the corner of a rectangular card lying on the table. The corner seen through the lens will only appear complete and continuous with the rest of the card when its tip is opposite the optical center.

In Fig. 33, A represents a lens so held that its optical

center is marked by the corner of the underlying card; B is a lens improperly held. The center first found may be marked with a speck of ink, the center of the other spectacle glass found in the same way, and the distance between them measured. If care is taken to hold the glass exactly level and the eye directly over it this method will give results accurate enough for most purposes.

The Apex of a Prism may be determined by viewing through the glass fine lines crossed at right angles, holding the prism so that its edge and supposed apex just touches one line at the point of intersection. When the real apex of the prism coincides with the intersection of the lines, the

FIG. 34. FIG. 35.

METHOD OF FINDING THE APEX OF A PRISM. (*After Maddox.*)

appearance presented is that shown in Fig. 34; when, however, the apex is to one side of the point of intersection, the line seen through the prism appears broken, as in Fig. 35. In this case the prism is to be rotated until the line appears continuous, when the point of intersection of the lines will mark the apex of the prism.

The strength of a prism may be expressed in two ways; either in degrees of the refracting angle, which is the angle forming the edge and separating the two refracting surfaces of the prism, or by means of some formula which denotes the power of the prism to turn a ray of light from its course. This power is usually expressed in degrees of the angle of deviation, which is the angle separating the course of a ray of light after having passed through the prism from that which it would have pursued had its course been unobstructed. The obvious advantage of the latter mode of expression, which gives directly the optical strength of the prism, over the former, which merely states the value of a physical angle from which the strength can be more or less accurately inferred, has called forth several suggestions for an improved method of numbering ophthalmological prisms. Dr. Edward Jackson has proposed, that in harmony with the mode of stating the value of angles which is commonly accepted in other departments of science, they be marked in degrees of their angles of deviation. With the idea of conforming their numeration to the dioptric system of numbering lenses, Mr. C. F. Prentice proposed to adopt as a unit that prism having the power necessary to produce one centimeter of deviation in the course of the ray after having passed through and the distance of one meter beyond the prism. Dr. S. M. Burnett proposes that this unit be called the prism diopter, and that the centimeter of deviation be measured upon a plane surface ; that is, upon a tangent of the arc whose radius is one meter Dr. W. S. Dennett prefers to call the unit a centrad, and to measure the deviation on the arc itself. Pending the general adoption of one of these proposals, it is sufficient for our purpose to note in regard to the two latter, that for ophthalmological prisms, which are of necessity weak, the difference between measuring the amount of deviation on an arc of given radius and on a tangent of that arc is so slight as to be of no moment. Prism diopters and centrads, therefore, though in the scientific aspect of the subject they represent distinct ideas, may be regarded as of equal value when speaking of the strength of prismatic spectacle glasses.

As the surgeon has a choice of three essentially different methods of numbering, so, also, he has at his command several modes of determining the strength of unknown prisms and may select that one which is simplest and involves least calculation for the numeration which he adopts. The refracting angle may be readily found by means of Table II, introduced when speaking of the prismatic equivalent of decentered lenses. The situation of the optical center is to be marked upon a spherical lens of convenient strength, and the prism to be tested superimposed. By viewing the corner of a card through these two glasses, as was directed in describing the method of finding the optical center, this center will be found to have been carried toward the base of the prism. The position of this apparent optical center is to be likewise marked upon the spherical lens, and its distance from the true one measured. In the left-hand column of Table II find the strength of the lens used, and on a level with this across the page the distance in millimeters between the true and apparent optical centers. At the head of the column in which this measurement is found will stand the strength of the prism with which the lens was combined, this strength being expressed in degrees of the refracting angle. For instance, if having combined an unknown prism with a $+ 7$. D. lens we find the apparent displacement of the optical center to be 4 mm., the table shows at a glance that the refracting angle of the prism tested had a value of $3°$.

The refracting angle may be directly measured by adapting the legs of a pair of compasses to the two refracting surfaces and then laying the compasses on an ordinary protractor. Various other mechanical contrivances have been invented for effecting the same purpose, one of the best of which is represented in Fig. 36. It consists of a bed-plate A, upon the front of which is affixed a degree-circle G, and hinged to A at H is the upper plate B held up by the spring M, not plainly shown because it is under B. The upright face-plate C stands at right angles to B. On top of C is the degree-circle E. The index-finger F with the lower part D D' is made of steel and

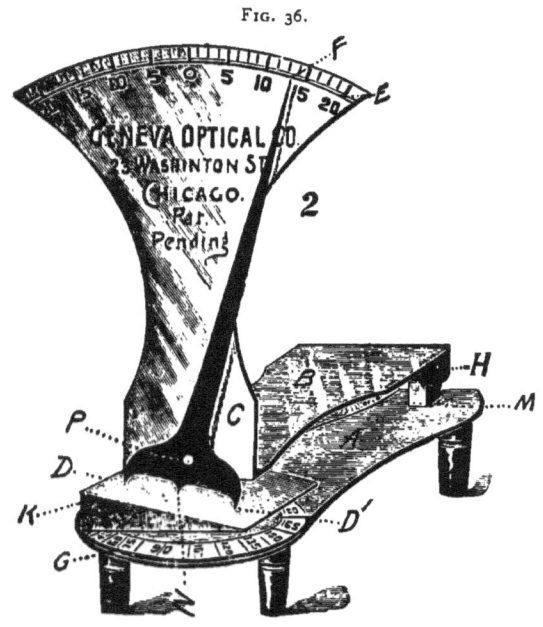

Fig. 36.

pivoted at P to swing easily over any portion of the dial plate. In measuring a prism, the position of the index finger F will be governed by the difference of the thickness of the lens at the points D and D', and the degrees of the refracting angle of the prism will be indicated on the scale E by the pointer F.

The surgeon is, however, very little concerned with the

INSPECTION AND ADJUSTMENT OF SPECTACLES. 81

refracting angles of the prisms, except as they are the basis of the old system of numbering, which will doubtless soon be superseded by one in which the number of the prism shall express in one of the ways mentioned above the power which that prism possesses of causing deviation in a ray of light. One of the simplest and most convenient devices for measuring this power is that suggested by Dr. Maddox. It consists of a strip of cardboard suspended horizontally on the wall on a level with the eyes of the observer. The upper border of the card (Fig. 37) is marked from right to left with a scale of degrees, or rather tangents of degrees, proper to the distance at which the prism is to be held from

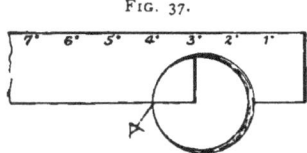

FIG. 37.

the card. In Table IV is given the distance from the right-hand border of the card of the mark for each degree of deviating angle. With the help of this table one may readily construct the scale, using column A if he elect to work at 6 feet, or column B if a 2-meter range be preferred.

To practice this method of prismetry, the glass to be tested is held at the proper distance from the card, its apex to the left, and its upper border just below the figures of the scale, as in Fig 37. The observer's eye being placed behind the prism, the right vertical border of the card appears displaced toward the observer's left and points upward to the number expressing the strength of the prism in degrees of the angle of deviation. During this maneuver care must be taken that the prism is held at precisely the distance from the card for which the scale of the latter is arranged; also that the apex of the prism points exactly to the left. This latter requirement may be secured by rotating the prism until the line of the bottom of the card appears unbroken, as at A, in Fig. 37. In adapting this method of prismetry to centrads or prism diopters, the scale at the top of the card should simply be laid off in centimeters, and the prism be held at the distance of one meter. Each cen-

TABLE IV.*

For Marking a Card in Tangents of Degrees at 6 Feet (Column A); or 2 Meters (Column B)					
	A	B		A	B
1°	1.25 in.	3.49 cm.	9°	11.4 in.	31.29 cm.
2°	2.5 "	6.98 "	10°	12.6 "	34.73 "
3°	3.7 "	10.467 "	11°	14.0 "	38.16 "
4°	5.0 "	13 95 "	12°	15.3 "	41.58 "
5°	6.3 "	17.43 "	13°	16 6 "	44.99 "
6°	7.57 "	20.9 "	14°	17.9 "	48.38 "
7°	8.84 "	24 37 "	15°	19.3 "	51.76 "
8°	10.12 "	27 83 "	16°	20.64 "	55.13 "

timeter that the right border of the card is apparently moved to the left on viewing it through the prism, will then represent one centrad, or one prism diopter.

Scratches, specks, bubbles, flaws, etc., in the glass will hardly escape detection if they are carefully looked for while the lens is held in different lights. Placing the glass against a dark background and allowing a bright light to fall obliquely upon it will perhaps bring them out as plainly as any other maneuver.

---

* From Maddox. "The Clinical Use of Prisms."

Irregularity of the Surface may be discovered by reflecting from that surface any object having regular outlines. The observer should stand facing a window, holding the lens against a dark background in his left hand, and pass a straight-edged piece of paper held in his right hand between his eyes and the lens. Two images of the paper will be reflected from the lens; one formed by each surface. Any irregularity of these surfaces will make the images appear broken, or with wavy outlines.

Adjusting Spectacle Frames.—It requires some little practice to enable one to tell at a glance just where such an irregularly shaped object as a spectacle frame has been wrongly bent; having found the error it is a more simple

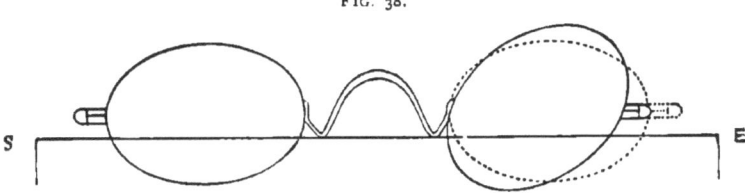

FIG. 38.

matter to correct it. For the latter purpose two small pliers are required. They should have narrow, but strong jaws; round in one pair, and square in the other. As found in the shops, the grasping surfaces of the jaws are generally roughened, but should be smoothed off with a file, lest they scar the gold when in use. A small, stout screw-driver with a point suited to the screws of spectacles will also be necessary.

Eye-wires are generally of such light material as to take their shape from the contained glass, and are therefore, not liable to become misshapen. Sometimes the long axis of an oval eye gets rotated within the eye-wire (Fig. 38), so that it no longer stands squarely across the face. By loosening the screw it can readily be re-adjusted. Abnormal crookedness about the bridge is best disclosed by placing a straight

FIG. 39.

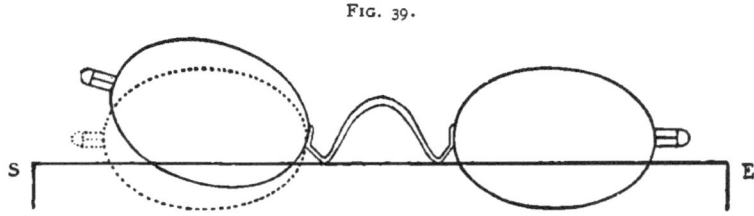

edge (indicated by the line S E in Figs. 38, 39, 41, 42, and 43) in such a position as to enable one to compare the two sides of the frame. If the bridge is bent at its junction

FIG. 40.

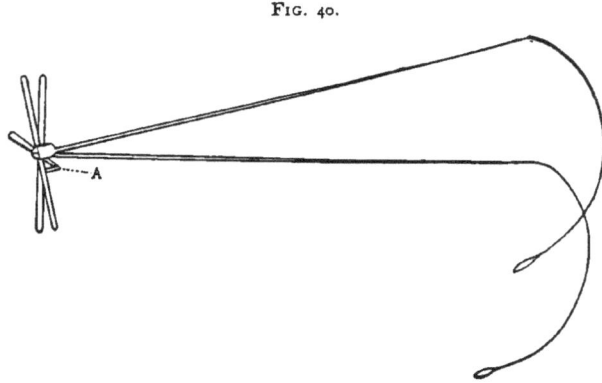

with the eye-wire a rotation results, looking very much like that just mentioned but dependent upon an entirely different fault (Fig. 39). It is readily corrected with pliers, or fingers.

INSPECTION AND ADJUSTMENT OF SPECTACLES.

FIG. 41.

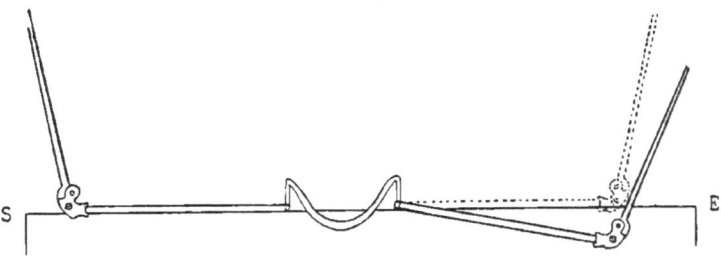

FIG. 42.

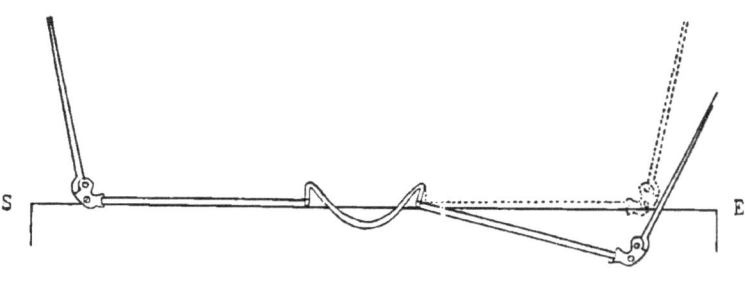

FIG. 43.

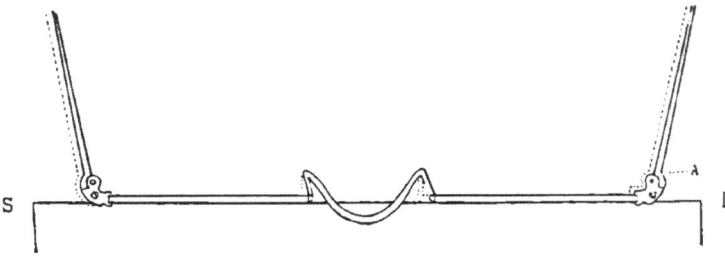

The planes of the glasses may cross each other (Fig. 40), in consequence of a twist in almost any part of the bridge, though the trouble is, usually, that the angle of the bridge at A is not of the same size as its fellow of the opposite side. The bridge is inclined, as shown in the cut, more to one glass than to the other. It requires application to the patient's face to determine which is the proper inclination, and in order that the glasses may be equalized at this and not at the improper one.

In Fig. 41, the bend is at the junction of the eye-wire with the bridge; rendering corresponding angles of the two sides of the frame unequal. The diagram shows the change necessary to correct the trouble. A similar fault is shown in Fig. 42. This appears at first sight to be just like the last; it is, however, a neighboring angle of the bridge which needs equalizing with its fellow.

In the frame represented in Fig. 43 the glasses lie in the same plane, but one of them is nearer the center of the bridge than the other, due to the fact that of the angles of the bridge which can be seen by viewing the frame in this position, the two which lie on one side of the curved portion are too much open, while the two on the other side are too little so. Of course, the bridge may be misshapen in any portion of its extent, but the illustrations given are sufficient to show the sort of faults one may expect.

Having rectified all want of symmetry in the "front," the defects in the fit of the temples can best be corrected by trying the frames on the patient's face. If on doing so it is found that their temples cut into the temples of the wearer, instead of just touching the skin as they should do, the trouble is obviously that the distance between the temples is too small, and they must be bent out at the hinges, so as to throw them, when open, further apart. This is done with the square-jawed pliers, seizing the wire close up to the hinge. When the opposite condition pertains, that is, when the distance between the temples is too great, leaving a space between each wire and the side of the wearer's head, they require to be bent in. To do this, take the end of each side in turn in the square-jawed pliers, in such a way that the edge of one jaw shall be in contact with the temple as close to the hinge as possible and the latter be held rigidly open. The temple may then be pressed in with the fingers, and will bend at the point where it is pressed against the edge of the pliers. If the latter are rightly placed this does not make an angle in the wire forming the temple, but simply alters the angle already formed at A in Fig. 43, by the expansion of the end of the temple to help form the hinge. Care must be taken that one temple is not bent out more than the other, or, as is apt to be the case, become so during use. When this happens the effect is quite different from what might be expected. The glass on the same side as the temple the more bent out will be brought closer to the eye, while its fellow will be carried further forward and the bridge will ride obliquely across the nose. To remedy this it is only necessary to equalize the divergence of the temples.

The curve of hook temples given them by the maker will rarely be found to fit comfortably behind the ear. As has been pointed out by Dr. Charles H. Thomas, the proper form for hook temples is a straight line from the hinge to the top of the ear, where a sharp curve should join this part of the temple to the easy curve which corresponds to the back of the ear (Fig. 40). Where the curve given the hook is too wide and is extended upon that part of the wire resting against the patient's temple, as shown by the dotted line in Fig. 44, there is a constant tendency of the spectacles to slide forward. The wire, moreover, touches the back of the ear for a short distance only, where its pressure is further increased by the fact of the whole temple being put upon the stretch and acting as a spring. Especially at first should the frames not fit too tightly, as the skin is then more easily irritated by the wire than when it becomes accustomed to its presence.

In persons whose ears stand out far from the head a certain ridge upon the cartilage of the ear is thrown into prominence. Since the curve of a hook temple is a regular one, it will rest upon this ridge and be very uncomfortable; indeed it may cut through the skin and into the cartilage.

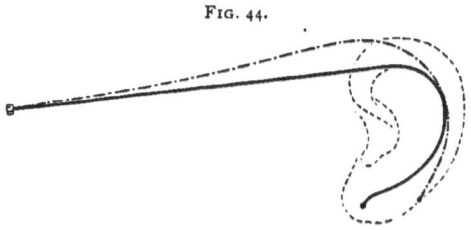

FIG. 44.

Under such circumstances the portion of the wire which is behind the ear should be made to follow every depression and elevation of the surface with which it is in contact; as it should in any case where the auricle is deformed or irregular in any way.

If one lens stands higher upon the face than the other, so that the patient looks through the upper part of one glass and the lower part of the other, it will be found that the temple on the side which stands the higher is turned down more than its fellow. It should be raised, or more frequently, its fellow should be lowered. The fault may lie in the bridge, as shown in Fig. 40, or in the end piece, or in the temple itself. In the first instance, bringing the lenses into the same plane removes the difficulty; in the second, take the end piece in the round-jawed pliers; the jaws being applied to its edges close up to the eye wire. Holding these pliers in the left hand, apply the square jaws of the other pliers to the surfaces of the end piece; when, by twisting the latter about its long axis, the temple may be turned down to any desired extent. Thus, the temple is not bent at all, but the end piece between the hinge and the eye wire. Nearly the same effect may be produced by bending the wire of the temple close up to the hinge. As was remarked before, in speaking of the facing

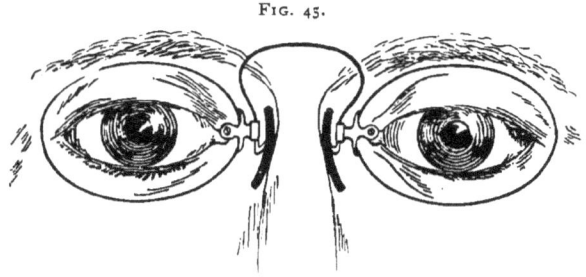

FIG. 45.

of the glasses, the effect of turning down both temples is not to make both lenses stand higher upon the face, but to make the glasses face more downward.

Sometimes when the glasses do not sit properly the trouble will be found to be not in the frames but in the wearer. A considerable amount of asymmetry of the two sides of the face is not uncommon. One ear or one eye may be higher than its fellow; either of which conditions will make the glasses seem awry, and render necessary a compensating asymmetry of their frames.

Adjustment of Eyeglasses. — The starting-point in adjusting eyeglasses is at the nose pieces, whose free surfaces should be made to conform accurately to the bones of the nose by which they are supported When received from the maker they are generally curved, presenting a convexity toward the nose. As the bones of the sides of the nose at the point where the guards are to rest are usually more or less convex also, the bearing obtained is a most insecure and uncomfortable one, as a glance at Fig. 45 will show. In Fig. 46 this glass is shown with its nose pieces properly adapted to the sides of the nose. Any conformation may be required, but that shown in Fig. 46 is the one most frequently needed. These changes in the

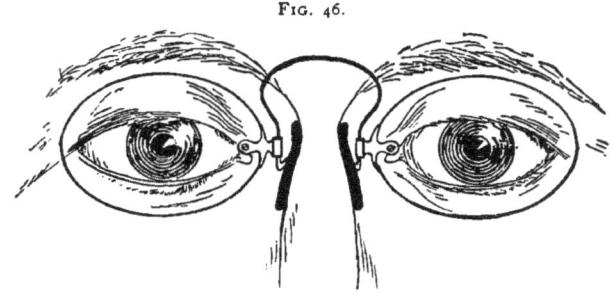

FIG. 46.

shape of the nose pieces are readily effected by means of the square-jawed pliers, especially if the so-called shell guards are used. The celluloid of which they are really made is, together with its gold backing, readily moulded into whatever shape is desired. When the guards are of cork, care must be taken that they are not scarred and broken by the pliers, and a special tool with a longitudinal groove in the jaws for grasping the sides of the nose pieces is here of service.

Having conformed the nose pieces to their bony support, the tension of the spring by which they are pressed against the sides of the nose is to be regulated, the object being to

INSPECTION AND ADJUSTMENT OF SPECTACLES. 91

have just sufficient force exerted to keep the guards securely in place. If the latter are properly fitted the amount of pressure necessary is not great. Though this pressure should be evenly distributed over the surfaces of the nose pieces, want of firmness in the "pinch" of their tops is particularly fatal, as the lower ends then become the principal support of the weight of the glasses, rendering them prone to topple forward and fall. To increase the tension of the spring, and consequently the pinch of the frames,

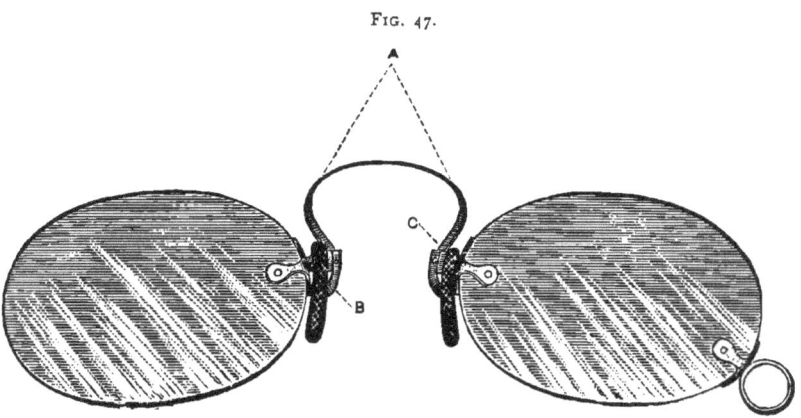

FIG. 47.

the curve of the spring included between the lines at A, in Fig. 47, should be made more arched and rounded. Conversely, the force of the spring is lessened by flattening this arch. Any alteration in the shape of the spring, however, while it does not, of course, change the shape of the nose pieces, does change the angle at which they are inclined to each other. For instance, if the spring be made more arched, the nose pieces are brought nearer together, but the bottoms are especially approached toward each other. When the spring is flattened the bottoms of the nose pieces are thrown proportionately farther apart than the tops. It follows that with each adjustment of the tension of the spring the inclination of the nose pieces must be rectified. This is easily accomplished by twisting the "foot" or support of the nose piece at B in Fig. 47. It will be readily seen, moreover, that the nose pieces must incline equally to a vertical plane passing through the center of the nose; otherwise the glasses will stand awry.

When the points mentioned have been properly adjusted, the long axis of one or both glasses may fail to stand squarely across the face as it should do. The remedy lies in an appropriate bend of the spring at the point C (Fig. 47). This also requires a slight re-adjustment of the inclination of the nose pieces to each other.

The distance between the centers of eyeglasses is determined (the distance between the nose pieces when in use being a fixed quantity) by the distance of the nose piece on each side from the center of the corresponding eye. The intercentral measurement may therefore be varied by varying the size of the eye used, and by altering the distance of the nose pieces from the edges of the lenses by an appropriate bend of the foot B, (Fig. 47). The distance of the glasses from the eye is controlled by the length of the foot B, and in the better grades of goods this part is made in two or three lengths.

The Care of Spectacles.—Spectacle frames will last longer and perform their function better if the wearer is instructed to exercise care in handling them. In putting them on and off, the hooks should be lifted from or into their position behind the ears; both hands being used, so as to avoid straining the temples widely apart or otherwise bending them. They should be folded together as little as possible, and when not in use should be laid in a safe place,

INSPECTION AND ADJUSTMENT OF SPECTACLES.

open, and resting on the edges of the lenses, to avoid scratching the surfaces of the latter. For cleansing them nothing is better than a piece of clean old linen, or, if very much soiled, a little ammonia and water may be used, except on cemented bifocal glasses. While cleansing, the frame should be grasped by the end piece and not by the bridge, and in replacing the glasses on the eyes care should be taken not to crush them against the lashes and thus soil the refracting surfaces at once. When cylindrical or prismatic glasses are worn, patients may return after a time with the statement that the spectacles are unsatisfactory, when the trouble will frequently be found to be due to bending of the frame; or a lens may have fallen out and been replaced upside down, or with the wrong edge inward. It is well to have such persons report periodically to have their glasses re-adjusted.

INDEX.

Adjustment of eyeglasses, 89
    of spectacles, 83
Airy, discoverer of astigmatism, 23
Alhazen, 19
Ancient glass, 17
Angle, deviating, of a prism, 78, 81
    refracting, 77, 79
Apex of a prism, finding the, 77
Assyrians, knowledge of lenses among, 18
Astigmatism, discovery of, 23
Asymmetry of the face, 89

Bar spring eyeglasses, 42
Bifocal glasses, 37
    invention of, 23
    varieties of, 39
Brewster, Sir David, 18, 23
Bridge, width of base of, 67
    relation of top of to plane of glasses, 52, 66
    height of, 65
Bridges, manufacture of, 34
    varieties of, 36

Care of spectacles, 92
Cemented bifocals, 39
Centering and decentering, 44
    normal lateral, 49
    normal vertical, 50
    of spectacles for constant use, 49
    near work, 49, 51
Centrad, 78
Chemical composition of glass, 29
Component parts of spectacles, 27
Conformation of nose pieces, 90
Crown glass, 29
Cylinder, finding the axis of, 74

Date of invention of spectacles, 20

Decentered lenses, 45
    prismatic effect of, 46
Deviating angle of a prism, 78, 81
Di Spina, Alessandro, 20
Discovery of astigmatism, 23
Distance between the pupils, 49, 61
Distance between temples, 86
    of the glasses from the eyes, 51

Emerald used by Nero, 18
Epicanthus, eyeglasses for, 43
Eyeglasses, advantages and disadvantages of, 26
    inspection and adjustment of, 70, 89
    prescription of, 69
    varieties of, 40
Extra front, 40
Eye wires, manufacture of, 34

Face, asymmetry of, 64, 68, 89
Facing of spectacles, 53
Focal length of lenses, 70
Frames (see spectacle frames)
Frameless spectacles, 34
Franklin, inventor of bifocal glasses, 23
Franklin glasses, 37

Geometrical center, 44
Glass, chemical composition of, 29
    ancient, 17
Ground bifocals, 39

Height of bridge, 65
Hook temples, 35, 87

Inspection and adjustment of spectacles, 70
Interpupillary distance, 49, 59, 61
Introduction, 17

INDEX.

Invention of spectacles, 20
Irregularity of surface of lenses, 83

Kepler, Johann, 22

Lateral centering, 49
Lathe for grinding lenses, 30
Lens, oldest known, 18
Lenses, decentered, 45
   finding focal length of, 70
   known to the ancients, 18
   material of, 28
   method of grinding, 30
   neutralization of spherical, 73
     cylindrical, 74
     sphero-cylindrical, 75
   proving the strength of, 70
   toric, 33, 57
   tilted, 52
Locating the optical center, 76
Lorgnettes, 26

Maddox pupil localizer, 63
Marking of gold spectacle frames, 27
Material of lenses, 28
Material of spectacle frames, 26

Natural position for spectacle bridge, 50
Nero, concave jewel used by, 18
Neutralization of spherical lenses, 73
   cylindrical lenses, 74
   sphero-cylindrical lenses, 75
Normal position of spectacles, 44
Nose pieces, conformation of, 90
Numeration of prisms, 78

Offset guard, 42
Oldest known lens, 18
Optician's lathe, 30
Optical center, 44
   locating the, 76

Pebble spectacles, 29
Periscopic glasses, 32, 56
Pince-nez (see eyeglasses)
Phacometers, 71
Plane of the glasses, relation to the visual axis, 52
Prism diopter, 78

Prism, finding the apex of, 77
   deviating angle of, 78, 81
   refracting angle of, 77, 79
Prisms, numeration of, 78
Prismatic effect of decentering, 46, 48
Prismetry, 77
Prescription blank for spectacles, 61
Prescription of eyeglasses, 69
Prescription of frames, 59
Proving the strength of lenses, 70
Principles of spectacle fitting, 44
Principal axis, 44
Pupil localizer, 63
Pupilometer, 65

Quizzing glasses, 26

Refracting angle of a prism, 77, 79
Rigid frame eyeglasses, 42
Rock crystal, 18, 29
Romans, knowledge of lenses among, 18
Rule for measuring frames, 59

Saddle bridge, 36
Salvinus Armatus, 20
Scratches, specks, flaws, etc., in glass, 82
Spectacle eyes, sizes of, 37
   shapes of, 37
   fitting, principles of, 44
   frames, adjustment of, 83
   material of, 26
   marking of gold, 27
   prescription of, 59
   rule for measuring, 59
Spectacles, component parts of, 27
   bifocal, 37
   care of, 92
   date of invention of, 20
   early references to, 20, 21
   for cosmetic effect, 43
   invention of bifocal, 23
   inspection and adjustment of, 70
   frameless, 34
   facing of, 53
   patterns of, 34
   pebble, 29
   periscopic, 32, 56

INDEX.

St. Jerome's eyeglasses, 21
Surface, irregularity of, 83

Temples, distance between, 86
    manufacture of, 34
    varieties of, 35
Tilted lenses, 52
"Tool" for grinding spherical lenses, 30
Tools for adjusting frames, 83

Toric lenses, 33, 57
Transparent glass found in Nineveh, 17
Trial frames, 64, 68

Vertical centering, 50
Visual axis, relation of plane of the glasses to, 52

Width of base of bridge, 67

CATALOGUE No. 7.   MARCH, 1892.

A CATALOGUE
OF
BOOKS FOR STUDENTS.
INCLUDING THE
? QUIZ-COMPENDS ?

CONTENTS.

	PAGE		PAGE
New Series of Manuals,	2,3,4,5	Obstetrics,	10
Anatomy,	6	Pathology, Histology,	11
Biology,	11	Pharmacy,	12
Chemistry,	6	Physical Diagnosis,	11
Children's Diseases,	7	Physiology,	11
Dentistry,	8	Practice of Medicine,	11, 12
Dictionaries,	8, 16	Prescription Books,	12
Eye Diseases,	9	? Quiz-Compends ?	14, 15
Electricity,	9	Skin Diseases,	12
Gynæcology,	10	Surgery and Bandaging,	13
Hygiene,	9	Therapeutics,	9
Materia Medica,	9	Urine and Urinary Organs,	13
Medical Jurisprudence,	10	Venereal Diseases,	13

PUBLISHED BY

P. BLAKISTON, SON & CO.,

Medical Booksellers, Importers and Publishers.

LARGE STOCK OF ALL STUDENTS' BOOKS, AT THE LOWEST PRICES.

1012 Walnut Street, Philadelphia.

*** For sale by all Booksellers, or any book will be sent by mail, postpaid, upon receipt of price. Catalogues of books on all branches of Medicine, Dentistry, Pharmacy, etc., supplied upon application.

Just Ready } 3000 Questions on Medical Subjects.
Price 10 cents

"An excellent Series of Manuals."—*Archives of Gynæcology.*

A NEW SERIES OF STUDENTS' MANUALS

On the various Branches of Medicine and Surgery.

Can be used by Students of any College.

Price of each, Handsome Cloth, $3.00. Full Leather, $3.50.

The object of this series is to furnish good manuals for the medical student, that will strike the medium between the compend on one hand and the prolix text-book on the other—to contain all that is necessary for the student, without embarrassing him with a flood of theory and involved statements. They have been prepared by well-known men, who have had large experience as teachers and writers, and who are, therefore, well informed as to the needs of the student.

Their mechanical execution is of the best—good type and paper, handsomely illustrated whenever illustrations are of use, and strongly bound in uniform style.

Each book is sold separately at a remarkably low price, and the immediate success of several of the volumes shows that the series has met with popular favor.

No. 1. SURGERY. 318 Illustrations.

Third Edition.

A Manual of the Practice of Surgery. By WM. J. WALSHAM, M.D., Asst. Surg. to, and Demonstrator of Surg. in, St. Bartholomew's Hospital, London, etc. 318 Illustrations.

Presents the introductory facts in Surgery in clear, precise language, and contains all the latest advances in Pathology, Antiseptics, etc.

"It aims to occupy a position midway between the pretentious manual and the cumbersome System of Surgery, and its general character may be summed up in one word—practical."—*The Medical Bulletin.*

"Walsham, besides being an excellent surgeon, is a teacher in its best sense, and having had very great experience in the preparation of candidates for examination, and their subsequent professional career, may be relied upon to have carried out his work successfully. Without following out in detail his arrangement, which is excellent, we can at once say that his book is an embodiment of modern ideas neatly strung together, with an amount of careful organization well suited to the candidate, and, indeed, to the practitioner."—*British Medical Journal.*

Price of each Book, Cloth, $3.00; Leather, $3.50.

No. 2. DISEASES OF WOMEN. 150 Illus.
NEW EDITION.

The Diseases of Women. Including Diseases of the Bladder and Urethra. By Dr. F. WINCKEL, Professor of Gynæcology and Director of the Royal University Clinic for Women, in Munich. Second Edition. Revised and Edited by Theophilus Parvin, M.D., Professor of Obstetrics and Diseases of Women and Children in Jefferson Medical College. 150 Engravings, most of which are original.

"The book will be a valuable one to physicians, and a safe and satisfactory one to put into the hands of students. It is issued in a neat and attractive form, and at a very reasonable price."—*Boston Medical and Surgical Journal.*

No. 3. OBSTETRICS. 227 Illustrations.

A Manual of Midwifery. By ALFRED LEWIS GALABIN, M.A., M.D., Obstetric Physician and Lecturer on Midwifery and the Diseases of Women at Guy's Hospital, London; Examiner in Midwifery to the Conjoint Examining Board of England, etc. With 227 Illus.

"This manual is one we can strongly recommend to all who desire to study the science as well as the practice of midwifery. Students at the present time not only are expected to know the principles of diagnosis, and the treatment of the various emergencies and complications that occur in the practice of midwifery, but find that the tendency is for examiners to ask more questions relating to the science of the subject than was the custom a few years ago. * * * The general standard of the manual is high; and wherever the science and practice of midwifery are well taught it will be regarded as one of the most important text-books on the subject."—*London Practitioner.*

No. 4. PHYSIOLOGY. Fifth Edition.
321 ILLUSTRATIONS AND A GLOSSARY.

A Manual of Physiology. By GERALD F. YEO, M.D., F.R.C.S., Professor of Physiology in King's College, London. 321 Illustrations and a Glossary of Terms. Fifth American from last English Edition, revised and improved. 758 pages.

This volume was specially prepared to furnish students with a new text-book of Physiology, elementary so far as to avoid theories which have not borne the test of time and such details of methods as are unnecessary for students in our medical colleges.

"The brief examination I have given it was so favorable that I placed it in the list of text-books recommended in the circular of the University Medical College."—*Prof. Lewis A. Stimson*, M.D., *37 East 33d Street, New York.*

Price of each Book, Cloth, $3.00; Leather, $3.50.

No. 5. DISEASES OF CHILDREN.

SECOND EDITION.

A Manual. By J. F. GOODHART, M.D., Phys. to the Evelina Hospital for Children; Asst. Phys. to Guy's Hospital, London. Second American Edition. Edited and Rearranged by LOUIS STARR, M.D., Clinical Prof. of Dis. of Children in the Hospital of the Univ. of Pennsylvania, and Physician to the Children's Hospital, Phila. Containing many new Prescriptions, a list of over 50 Formulæ, conforming to the U. S. Pharmacopœia, and Directions for making Artificial Human Milk, for the Artificial Digestion of Milk, etc. Illus.

"The merits of the book are many. Aside from the praiseworthy work of the printer and binder, which gives us a print and page that delights the eye, there is the added charm of a style of writing that is not wearisome, that makes its statements clearly and forcibly, and that knows when to stop when it has said enough. The insertion of typical temperature charts certainly enhances the value of the book. It is rare, too, to find in any text-book so many topics treated of. All the rarer and out-of-the-way diseases are given consideration. This we commend. It makes the work valuable."—*Archives of Pediatrics, July, 1890.*

"The author has avoided the not uncommon error of writing a book on general medicine and labeling it 'Diseases of Children,' but has steadily kept in view the diseases which seemed to be incidental to childhood, or such points in disease as appear to be so peculiar to or pronounced in children as to justify insistence upon them. * * * A safe and reliable guide, and in many ways admirably adapted to the wants of the student and practitioner."—*American Journal of Medical Science.*

"Thoroughly individual, original and earnest, the work evidently of a close observer and an independent thinker, this book, though small, as a handbook or compendium is by no means made up of bare outlines or standard facts."—*The Therapeutic Gazette.*

"As it is said of some men, so it might be said of some books, that they are 'born to greatness.' This new volume has, we believe, a mission, particularly in the hands of the younger members of the profession. In these days of prolixity in medical literature, it is refreshing to meet with an author who knows both what to say and when he has said it. The work of Dr. Goodhart (admirably conformed, by Dr. Starr, to meet American requirements) is the nearest approach to clinical teaching without the actual presence of clinical material that we have yet seen."—*New York Medical Record.*

Price of each Book, Cloth, $3.00; Leather, $3.50.

No. 6. PRACTICAL THERAPEUTICS.
FOURTH EDITION, WITH AN INDEX OF DISEASES.

Practical Therapeutics, considered with reference to Articles of the Materia Medica. Containing, also, an Index of Diseases, with a list of the Medicines applicable as Remedies. By EDWARD JOHN WARING, M.D., F.R.C.P. Fourth Edition. Rewritten and Revised by DUDLEY W. BUXTON, M.D., Asst. to the Prof. of Medicine at University College Hospital.

"We wish a copy could be put in the hands of every Student or Practitioner in the country. In our estimation, it is the best book of the kind ever written."—*N. Y. Medical Journal.*

"Dr. Waring's Therapeutics has long been known as one of the most thorough and valuable of medical works. The amount of actual intellectual labor it represents is immense. . . . An index of diseases, with the remedies appropriate for their treatment, closes the volume."—*Boston Medical and Surgical Reporter.*

"The plan of this work is an admirable one, and one well calculated to meet the wants of busy practitioners. There is a remarkable amount of information, accompanied with judicious comments, imparted in a concise yet agreeable style."—*Medical Record.*

No. 7. MEDICAL JURISPRUDENCE AND TOXICOLOGY.
THIRD REVISED EDITION.

By JOHN J. REESE, M.D., Professor of Medical Jurisprudence and Toxicology in the University of Pennsylvania; President of the Medical Jurisprudence Society of Phila.; Third Edition, Revised and Enlarged.

"This admirable text-book."—*Amer. Jour. of Med. Sciences.*

"We lay this volume aside, after a careful perusal of its pages, with the profound impression that it should be in the hands of every doctor and lawyer. It fully meets the wants of all students. . . . He has succeeded in admirably condensing into a handy volume all the essential points."—*Cincinnati Lancet and Clinic.*

"The book before us will, we think, be found to answer the expectations of the student or practitioner seeking a manual of jurisprudence, and the call for a second edition is a flattering testimony to the value of the author's present effort. The medical portion of this volume seems to be uniformly excellent, leaving little for adverse criticism. The information on the subject matter treated has been carefully compiled, in accordance with recent knowledge. The toxicological portion appears specially excellent. Of that portion of the work treating of the legal relations of the practitioner and medical witness, we can express a generally favorable verdict."—*Physician and Surgeon, Ann Arbor, Mich.*

Price of each Book, Cloth, $3,00; Leather, $3.50.

ANATOMY.

Macalister's Human Anatomy. 816 Illustrations. A new Text-book for Students and Practitioners, Systematic and Topographical, including the Embryology, Histology and Morphology of Man. With special reference to the requirements of Practical Surgery and Medicine. With 816 Illustrations, 400 of which are original. Octavo. Cloth, 7.50; Leather, 8.50

Ballou's Veterinary Anatomy and Physiology. Illustrated. By Wm. R. Ballou, M.D., Professor of Equine Anatomy at New York College of Veterinary Surgeons. 29 graphic Illustrations. 12mo. Cloth, 1.00; Interleaved for notes, 1.25

Holden's Anatomy. A manual of Dissection of the Human Body. Fifth Edition. Enlarged, with Marginal References and over 200 Illustrations. Octavo.

Bound in Oilcloth, for the Dissecting Room, $4.50.

"No student of Anatomy can take up this book without being pleased and instructed. Its Diagrams are original, striking and suggestive, giving more at a glance than pages of text description. * * * The text matches the illustrations in directness of practical application and clearness of detail."—*New York Medical Record.*

Holden's Human Osteology. Comprising a Description of the Bones, with Colored Delineations of the Attachments of the Muscles. The General and Microscopical Structure of Bone and its Development. With Lithographic Plates and Numerous Illustrations. Seventh Edition. 8vo. Cloth, 6.00

Holden's Landmarks, Medical and Surgical. 4th ed. Clo., 1.25

Heath's Practical Anatomy. Sixth London Edition. 24 Colored Plates, and nearly 300 other Illustrations. Cloth, 5.00

Potter's Compend of Anatomy. Fifth Edition. Enlarged. 16 Lithographic Plates. 117 Illustrations. *See Page 14.*
Cloth, 1.00; Interleaved for Notes, 1.25

CHEMISTRY.

Bartley's Medical Chemistry. Second Edition. A text-book prepared specially for Medical, Pharmaceutical and Dental Students. With 50 Illustrations, Plate of Absorption Spectra and Glossary of Chemical Terms. Revised and Enlarged. Cloth, 2.50

Trimble. Practical and Analytical Chemistry. A Course in Chemical Analysis, by Henry Trimble, Prof. of Analytical Chemistry in the Phila. College of Pharmacy. Illustrated. Fourth Edition, Enlarged. 8vo. Cloth, 1.50

☞ *See pages 2 to 5 for list of Students' Manuals.*

Chemistry:—Continued.

Bloxam's Chemistry, Inorganic and Organic, with Experiments. Seventh Edition. Enlarged and Rewritten. 281 Illustrations.
Cloth, 4.50; Leather, 5.50

Richter's Inorganic Chemistry. A text-book for Students. Third American, from Fifth German Edition. Translated by Prof. Edgar F. Smith, PH.D. 89 Wood Engravings and Colored Plate of Spectra. Cloth, 2.00

Richter's Organic Chemistry, or Chemistry of the Carbon Compounds. Illustrated. Second Edition. Cloth, 4.50

Symonds. Manual of Chemistry, for the special use of Medical Students. By BRANDRETH SYMONDS, A.M., M.D., Asst. Physician Roosevelt Hospital, Out-Patient Department; Attending Physician Northwestern Dispensary, New York. 12mo.
Cloth, 2.00

Leffmann's Compend of Chemistry. Inorganic and Organic. Including Urinary Analysis. Third Edition. Revised.
Cloth, 1.00; Interleaved for Notes, 1.25

Leffmann and Beam. Progressive Exercises in Practical Chemistry. 12mo. Illustrated. Cloth, 1.00

Muter. Practical and Analytical Chemistry. Fourth Edition. Revised, to meet the requirements of American Medical Colleges, by Prof. C. C. Hamilton. Illustrated. Cloth, 2.00

Holland. The Urine, Common Poisons, and Milk Analysis, Chemical and Microscopical. For Laboratory Use. Fourth Edition, Enlarged. Illustrated. Cloth, 1.00

Van Nüys. Urine Analysis. Illus. Cloth, 2.00

Wolff's Applied Medical Chemistry. By Lawrence Wolff, M.D., Dem. of Chemistry in Jefferson Medical College. Clo., 1.00

CHILDREN.

Goodhart and Starr. The Diseases of Children. Second Edition. By J. F. Goodhart, M.D., Physician to the Evelina Hospital for Children; Assistant Physician to Guy's Hospital, London. Revised and Edited by Louis Starr, M.D., Clinical Professor of Diseases of Children in the Hospital of the University of Pennsylvania; Physician to the Children's Hospital, Philadelphia. Containing many Prescriptions and Formulæ, conforming to the U. S. Pharmacopœia, Directions for making Artificial Human Milk, for the Artificial Digestion of Milk, etc. Illustrated. Cloth, 3.00; Leather, 3.50

Hatfield. Diseases of Children. By M. P. Hatfield, M.D., Professor of Diseases of Children, Chicago Medical College. Colored Plate. 12mo. Cloth, 1.00; Interleaved, 1.25

☞ *See pages 14 and 15 for list of ? Quiz-Compends!*

8   STUDENTS' TEXT-BOOKS AND MANUALS.

Children:—Continued.

Starr. Diseases of the Digestive Organs in Infancy and Childhood. With chapters on the Investigation of Disease, and on the General Management of Children. By Louis Starr, M.D., Clinical Professor of Diseases of Children in the University of Pennsylvania. Illus. Second Edition.   Cloth, 2.25

DENTISTRY.

Fillebrown. Operative Dentistry. 330 Illus.   Cloth, 2.50
Flagg's Plastics and Plastic Filling.   4th Ed.   Cloth, 4.00
Gorgas. Dental Medicine. A Manual of Materia Medica and Therapeutics. Fourth Edition.   Cloth, 3.50
Harris. Principles and Practice of Dentistry. Including Anatomy, Physiology, Pathology, Therapeutics, Dental Surgery and Mechanism. Twelfth Edition. Revised and enlarged by Professor Gorgas. 1028 Illustrations. Cloth, 7.00; Leather, 8.00
Richardson's Mechanical Dentistry. Fifth Edition. 569 Illustrations. 8vo.   Cloth, 4.50; Leather, 5.50
Sewill. Dental Surgery. 200 Illustrations. 3d Ed. Clo., 3.00
Taft's Operative Dentistry. Dental Students and Practitioners. Fourth Edition. 100 Illustrations.   Cloth, 4.25; Leather, 5.00
Talbot. Irregularities of the Teeth, and their Treatment. Illustrated. 8vo. Second Edition.   Cloth, 3.00
Tomes' Dental Anatomy. Third Ed.   191 Illus.   Cloth, 4.00
Tomes' Dental Surgery. 3d Edition. Revised. 292 Illus. 772 Pages.   Cloth, 5.00
Warren. Compend of Dental Pathology and Dental Medicine. Illustrated.   Cloth, 1.00; Interleaved, 1.25

DICTIONARIES.

Gould's New Medical Dictionary. Containing the Definition and Pronunciation of all words in Medicine, with many useful Tables etc. ½ Dark Leather, 3.25; ½ Mor., Thumb Index 4.25
Harris' Dictionary of Dentistry. Fifth Edition. Completely revised and brought up to date by Prof. Gorgas.
Cloth, 5.00; Leather, 6.00
Cleaveland's Pronouncing Pocket Medical Lexicon. 31st Edition. Giving correct Pronunciation and Definition. Very small pocket size. Cloth, red edges .75; pocket-book style, 1.00
Longley's Pocket Dictionary. The Student's Medical Lexicon, giving Definition and Pronunciation of all Terms used in Medicine, with an Appendix giving Poisons and Their Antidotes, Abbreviations used in Prescriptions, Metric Scale of Doses, etc. 24mo.   Cloth, 1.00; pocket-book style, 1.25

☛ *See pages 2 to 5 for list of Students' Manuals.*

EYE.

Hartridge on Refraction. 5th Edition. Illus. Cloth, 2.00

Hartridge on the Ophthalmoscope. Illustrated. Cloth, 1.50

Meyer. Diseases of the Eye. A complete Manual for Students and Physicians. 270 Illustrations and two Colored Plates. 8vo. Cloth, 4.50; Leather, 5.50

Swanzy. Diseases of the Eye and their Treatment. 158 Illustrations. Fourth Edition. Cloth, 3.00

Fox and Gould. Compend of Diseases of the Eye and Refraction. 2d Ed. Enlarged. 71 Illus. 39 Formulæ.
Cloth, 1.00; Interleaved for Notes, 1.25

ELECTRICITY.

Bigelow. Plain Talks on Medical Electricity and Batteries. Illustrated. With a Glossary of Electrical Terms. Cloth, 1.00

Mason's Compend of Medical and Surgical Electricity. With numerous Illustrations. 12mo. Cloth, 1.00

HYGIENE.

Parkes' (Ed. A.) Practical Hygiene. Seventh Edition, enlarged. Illustrated. 8vo. Cloth, 4.50

Parkes' (L. C.) Manual of Hygiene and Public Health. Second Edition. 12mo. Cloth, 2.50

Wilson's Handbook of Hygiene and Sanitary Science. Seventh Edition. Revised and Illustrated. *In Press.*

MATERIA MEDICA AND THERAPEUTICS.

Potter's Compend of Materia Medica, Therapeutics and Prescription Writing. Fifth Edition, revised and improved. *See Page 15.* Cloth, 1.00; Interleaved for Notes, 1.25

Biddle's Materia Medica. Eleventh Edition. By the late John B. Biddle, M.D., Prof. of Materia Medica in Jefferson College, Philadelphia. Revised by Clement Biddle, M.D., and Henry Morris, M.D. 8vo., illustrated. Cloth, 4.25; Leather, 5.00

Potter. Handbook of Materia Medica, Pharmacy and Therapeutics. Including Action of Medicines, Special Therapeutics, Pharmacology, etc. By Saml. O. L. Potter, M.D., M.R.C.P. (Lond.), Professor of the Practice of Medicine in Cooper Medical College, San Francisco. Third Revised and Enlarged Edition. 8vo. Cloth, 4.00; Leather, 5.00

Waring. Therapeutics. With an Index of Diseases and Remedies. 4th Edition. Revised. Cloth, 3.00; Leather, 3.50

☞ *See pages 14 and 15 for list of ? Quiz-Compends ?*

MEDICAL JURISPRUDENCE.

Reese. A Text-book of Medical Jurisprudence and Toxicology. By John J. Reese, M.D., Professor of Medical Jurisprudence and Toxicology in the Medical Department of the University of Pennsylvania; President of the Medical Jurisprudence Society of Philadelphia; Physician to St. Joseph's Hospital; Corresponding Member of The New York Medico-legal Society. Third Edition. Cloth, 3.00; Leather, 3.50

OBSTETRICS AND GYNÆCOLOGY.

Davis. A Manual of Obstetrics. By Edw. P. Davis, Demonstrator of Obstetrics, Jefferson Medical College, Philadelphia. Colored Plates, and 130 other Illustrations. 12mo. Cloth, 2.00

Byford. Diseases of Women. The Practice of Medicine and Surgery, as applied to the Diseases and Accidents Incident to Women. By W. H. Byford, A.M., M.D., Professor of Gynæcology in Rush Medical College and of Obstetrics in the Woman's Medical College, etc., and Henry T. Byford, M.D., Surgeon to the Woman's Hospital of Chicago. Fourth Edition. Revised and Enlarged. 306 Illustrations, over 100 of which are original. Octavo. 832 pages. Cloth, 5.00; Leather, 6.00

Cazeaux and Tarnier's Midwifery. With Appendix, by Mundé. The Theory and Practice of Obstetrics; including the Diseases of Pregnancy and Parturition, Obstetrical Operations, etc. Eighth American, from the Eighth French and First Italian Edition. Edited by Robert J. Hess, M.D., Physician to the Northern Dispensary, Philadelphia, with an appendix by Paul F. Mundé, M.D., Professor of Gynæcology at the N. Y. Polyclinic. Illustrated by Chromo-Lithographs, and other Full-page Plates, seven of which are beautifully colored, and numerous Wood Engravings. One Vol., 8vo. Cloth, 5.00; Leather, 6.00

Lewers' Diseases of Women. A Practical Text-Book. 139 Illustrations. Second Edition. Cloth, 2.50

Parvin's Winckel's Diseases of Women. Second Edition. Including a Section on Diseases of the Bladder and Urethra. 150 Illus. Revised. *See page 3.* Cloth, 3.00; Leather, 3.50

Morris. Compend of Gynæcology. Illustrated. Cloth, 1.00

Winckel's Obstetrics. A Text-book on Midwifery, including the Diseases of Childbed. By Dr. F. Winckel, Professor of Gynæcology, and Director of the Royal University Clinic for Women, in Munich. Authorized Translation, by J. Clifton Edgar, M.D., Lecturer on Obstetrics, University Medical College, New York, with nearly 200 handsome illustrations, the majority of which are original. 8vo. Cloth, 6.00; Leather, 7.00

Landis' Compend of Obstetrics. Illustrated. 4th edition, enlarged. Cloth, 1.00; Interleaved for Notes, 1.25

Galabin's Midwifery. By A. Lewis Galabin, M.D., F.R.C.P. 227 Illustrations. *See page 3.* Cloth, 3.00; Leather, 3.50

☞ *See pages 2 to 5 for list of New Manuals.*

PATHOLOGY. HISTOLOGY. BIOLOGY.

Bowlby. Surgical Pathology and Morbid Anatomy, for Students. 135 Illustrations. 12mo. Cloth, 2.00

Davis' Elementary Biology. Illustrated. Cloth, 4.00

Gilliam's Essentials of Pathology. A Handbook for Students. 47 Illustrations. 12mo. Cloth, 2.00

*** The object of this book is to unfold to the beginner the fundamentals of pathology in a plain, practical way, and by bringing them within easy comprehension to increase his interest in the study of the subject.

Gibbes' Practical Histology and Pathology. Third Edition. Enlarged. 12mo. Cloth, 1.75

Virchow's Post-Mortem Examinations. 3d Ed. Cloth, 1.00

PHYSICAL DIAGNOSIS.

Fenwick. Student's Guide to Physical Diagnosis. 7th Edition. 117 Illustrations. 12mo. Cloth, 2.25

Tyson's Student's Handbook of Physical Diagnosis. Illustrated. 12mo. Cloth, 1.25

PHYSIOLOGY.

Yeo's Physiology. Fifth Edition. The most Popular Students' Book. By Gerald F. Yeo, M.D., F.R.C.S., Professor of Physiology in King's College, London. Small Octavo. 758 pages. 321 carefully printed Illustrations. With a Full Glossary and Index. *See Page 3.* Cloth, 3.00; Leather, 3.50

Brubaker's Compend of Physiology. Illustrated. Sixth Edition. Cloth, 1.00; Interleaved for Notes, 1.25

Stirling. Practical Physiology, including Chemical and Experimental Physiology. 142 Illustrations. Cloth, 2.25

Kirke's Physiology. New 12th Ed. Thoroughly Revised and Enlarged. 502 Illustrations. Cloth, 4.00; Leather, 5.00

Landois' Human Physiology. Including Histology and Microscopical Anatomy, and with special reference to Practical Medicine. Fourth Edition. Translated and Edited by Prof. Stirling. 845 Illustrations. Cloth, 7.00; Leather, 8.00

" With this Text-book at his command, no student could fail in his examination."—*Lancet.*

Sanderson's Physiological Laboratory. Being Practical Exercises for the Student. 350 Illustrations. 8vo. Cloth, 5.00

PRACTICE.

Taylor. Practice of Medicine. A Manual. By Frederick Taylor, M.D., Physician to, and Lecturer on Medicine at, Guy's Hospital, London; Physician to Evelina Hospital for Sick Children, and Examiner in Materia Medica and Pharmaceutical Chemistry, University of London. Cloth, 4.00; Leather, 5.00

☞ *See pages 14 and 15 for list of ? Quiz-Compends ?*

12   STUDENTS' TEXT-BOOKS AND MANUALS.

Practice:—Continued.

Roberts' Practice. New Revised Edition. A Handbook of the Theory and Practice of Medicine. By Frederick T. Roberts, M.D.; M.R.C.P., Professor of Clinical Medicine and Therapeutics in University College Hospital, London. Seventh Edition. Octavo. Cloth, 5.50; Sheep, 6.50

Hughes. Compend of the Practice of Medicine. 4th Edition. Two parts, each, Cloth, 1.00; Interleaved for Notes, 1.25

PART I.—Continued, Eruptive and Periodical Fevers, Diseases of the Stomach, Intestines, Peritoneum, Biliary Passages, Liver, Kidneys, etc., and General Diseases, etc.

PART II.—Diseases of the Respiratory System, Circulatory System and Nervous System; Diseases of the Blood, etc.

Physicians' Edition. Fourth Edition. Including a Section on Skin Diseases. With Index. 1 vol. Full Morocco, Gilt, 2.50

From John A. Robinson, M.D., Assistant to Chair of Clinical Medicine, now Lecturer on Materia Medica, Rush Medical College, Chicago.

"Meets with my hearty approbation as a substitute for the ordinary note books almost universally used by medical students. It is concise, accurate, well arranged and lucid, . . . just the thing for students to use while studying physical diagnosis and the more practical departments of medicine."

PRESCRIPTION BOOKS.

Wythe's Dose and Symptom Book. Containing the Doses and Uses of all the principal Articles of the Materia Medica, etc. Seventeenth Edition. Completely Revised and Rewritten. *Just Ready.* 32mo. Cloth, 1.00; Pocket-book style, 1.25

Pereira's Physician's Prescription Book. Containing Lists of Terms, Phrases, Contractions and Abbreviations used in Prescriptions Explanatory Notes, Grammatical Construction of Prescriptions, etc., etc. By Professor Jonathan Pereira, M.D. Sixteenth Edition. 32mo. Cloth, 1.00; Pocket-book style, 1.25

PHARMACY.

Stewart's Compend of Pharmacy. Based upon Remington's Text-Book of Pharmacy. Third Edition, Revised. With new Tables, Index, Etc. Cloth, 1.00; Interleaved for Notes, 1.25

Robinson. Latin Grammar of Pharmacy and Medicine. By H. D. Robinson, PH.D., Professor of Latin Language and Literature, University of Kansas, Lawrence. With an Introduction by L. E. Sayre, PH.G., Professor of Pharmacy in, and Dean of, the Dept. of Pharmacy, University of Kansas. 12mo. Cloth, 2.00

SKIN DISEASES.

Anderson, (McCall) Skin Diseases. A complete Text-Book, with Colored Plates and numerous Wood Engravings. 8vo. Cloth, 4.50; Leather, 5.50

Van Harlingen on Skin Diseases. A Handbook of the Diseases of the Skin, their Diagnosis and Treatment (arranged alphabetically). By Arthur Van Harlingen, M.D., Clinical Lecturer on Dermatology, Jefferson Medical College; Prof. of Diseases of the Skin in the Philadelphia Polyclinic. 2d Edition. Enlarged. With colored and other plates and illustrations. 12mo. Cloth, 2.50

☞ *See pages 2 to 5 for list of New Manuals.*

SURGERY AND BANDAGING.

Moullin's Surgery, A new Text-Book. 500 Illustrations (some colored), 200 of which are original.
Cloth, net 7.00; Leather, net 8.00

Jacobson. Operations in Surgery. A Systematic Handbook for Physicians, Students and Hospital Surgeons. By W. H. A. Jacobson, B.A., Oxon. F.R.C.S. Eng.; Ass't Surgeon Guy's Hospital; Surgeon at Royal Hospital for Children and Women, etc. 199 Illustrations. 1006 pages. 8vo. Cloth. 5.00; Leather, 6.00

Heath's Minor Surgery, and Bandaging. Ninth Edition. 142 Illustrations. 60 Formulæ and Diet Lists. Cloth, 2.00

Horwitz's Compend of Surgery, Minor Surgery and Bandaging, Amputations, Fractures, Dislocations, Surgical Diseases, and the Latest Antiseptic Rules, etc., with Differential Diagnosis and Treatment. By ORVILLE HORWITZ, B.S., M.D., Demonstrator of Surgery, Jefferson Medical College. 4th edition. Enlarged and Rearranged. 136 Illustrations and 84 Formulæ. 12mo. Cloth, 1.00; Interleaved for the addition of Notes, 1.25
₂ The new Section on Bandaging and Surgical Dressings, consists of 32 Pages and 41 Illustrations. Every Bandage of any importance is figured. This, with the Section on Ligation of Arteries, forms an ample Text-book for the Surgical Laboratory.

Walsham. Manual of Practical Surgery. Third Edition. By WM. J. WALSHAM, M.D., F.R.C.S., Asst. Surg. to, and Dem. of Practical Surg. in, St. Bartholomew's Hospital; Surgeon to Metropolitan Free Hospital, London. With 318 Engravings. *See Page 2.* Cloth, 3.00; Leather, 3.50

URINE, URINARY ORGANS, ETC.

Holland. The Urine, and Common Poisons and The Milk. Chemical and Microscopical, for Laboratory Use. Illustrated. Fourth Edition. 12mo. Interleaved. Cloth, 1.00

Ralfe. Kidney Diseases and Urinary Derangements. 42 Illustrations. 12mo. 572 pages. Cloth, 2.75

Marshall and Smith. On the Urine. The Chemical Analysis of the Urine. By John Marshall, M.D., Chemical Laboratory, Univ. of Penna; and Prof. E. F. Smith, PH.D. Col. Plates. Cloth, 1.00

Tyson. On the Urine. A Practical Guide to the Examination of Urine. With Colored Plates and Wood Engravings. 7th Ed. Enlarged. 12mo. Cloth, 1.50

Van Nüys, Urine Analysis. Illus. Cloth, 2.00

VENEREAL DISEASES.

Hill and Cooper. Student's Manual of Venereal Diseases, with Formulæ. Fourth Edition. 12mo. Cloth, 1.00

☞ *See pages 14 and 15 for list of ? Quiz-Compends ?*

NEW AND REVISED EDITIONS.

? QUIZ-COMPENDS ?

The Best Compends for Students' Use in the Quiz Class, and when Preparing for Examinations.

Compiled in accordance with the latest teachings of prominent lecturers and the most popular Text-books.

They form a most complete, practical and exhaustive set of manuals, containing information nowhere else collected in such a condensed, practical shape. Thoroughly up to the times in every respect, containing many new prescriptions and formulæ, and over two hundred and fifty illustrations, many of which have been drawn and engraved specially for this series. The authors have had large experience as quiz-masters and attachés of colleges, with exceptional opportunities for noting the most recent advances and methods.

Cloth, each $1.00.   Interleaved for Notes, $1.25.

No. 1. HUMAN ANATOMY, "Based upon Gray." Fifth Enlarged Edition, including Visceral Anatomy, formerly published separately. 16 Lithograph Plates, New Tables and 117 other Illustrations. By SAMUEL O. L. POTTER, M.A., M.D., M.R.C.P. (Lond.,) late A. A. Surgeon U. S. Army. Professor of Practice, Cooper Medical College, San Francisco.

Nos. 2 and 3. PRACTICE OF MEDICINE. Fourth Edition. By DANIEL E. HUGHES, M.D., Demonstrator of Clinical Medicine in Jefferson Medical College, Philadelphia. In two parts.

PART I.—Continued, Eruptive and Periodical Fevers, Diseases of the Stomach, Intestines, Peritoneum, Biliary Passages, Liver, Kidneys, etc. (including Tests for Urine), General Diseases, etc.

PART II.—Diseases of the Respiratory System (including Physical Diagnosis), Circulatory System and Nervous System; Diseases of the Blood, etc.

* These little books can be regarded as a full set of notes upon the Practice of Medicine, containing the Synonyms, Definitions, Causes, Symptoms, Prognosis, Diagnosis, Treatment, etc., of each disease, and including a number of prescriptions hitherto unpublished.

No. 4. PHYSIOLOGY, including Embryology. Sixth Edition. By ALBERT P. BRUBAKER, M.D., Prof. of Physiology, Penn'a College of Dental Surgery; Demonstrator of Physiology in Jefferson Medical College, Philadelphia. Revised, Enlarged, with new Illustrations.

No. 5. OBSTETRICS. Illustrated. Fourth Edition. By HENRY G. LANDIS, M.D., Prof. of Obstetrics and Diseases of Women, in Starling Medical College, Columbus, O. Revised Edition. New Illustrations.

BLAKISTON'S ? QUIZ-COMPENDS ?

No. 6. MATERIA MEDICA, THERAPEUTICS AND PRESCRIPTION WRITING. Fifth Revised Edition. With especial Reference to the Physiological Action of Drugs, and a complete article on Prescription Writing. Based on the Last Revision of the U. S. Pharmacopœia, and including many unofficinal remedies. By SAMUEL O. L. POTTER, M.A., M.D., M.R.C.P. (Lond.,) late A. A. Surg. U. S. Army; Prof. of Practice, Cooper Medical College, San Francisco. Improved and Enlarged, with Index.

No. 7. GYNÆCOLOGY. A Compend of Diseases of Women. By HENRY MORRIS, M.D., Demonstrator of Obstetrics, Jefferson Medical College, Philadelphia. 45 Illustrations.

No. 8. DISEASES OF THE EYE AND REFRACTION, including Treatment and Surgery. By L. WEBSTER FOX, M.D., Chief Clinical Assistant Ophthalmological Dept., Jefferson Medical College, etc., and GEO. M. GOULD, M.D. 71 Illustrations, 39 Formulæ. Second Enlarged and Improved Edition. Index.

No. 9. SURGERY, Minor Surgery and Bandaging. Illustrated. Fourth Edition. Including Fractures, Wounds, Dislocations, Sprains, Amputations and other operations; Inflammation, Suppuration, Ulcers, Syphilis, Tumors, Shock, etc. Diseases of the Spine, Ear, Bladder, Testicles, Anus, and other Surgical Diseases. By ORVILLE HORWITZ, A.M., M.D., Demonstrator of Surgery, Jefferson Medical College. Revised and Enlarged. 84 Formulæ and 136 Illustrations.

No. 10. CHEMISTRY. Inorganic and Organic. For Medical and Dental Students. Including Urinary Analysis and Medical Chemistry. By HENRY LEFFMANN, M.D., Prof. of Chemistry in Penn'a College of Dental Surgery, Phila. Third Edition, Revised and Rewritten, with Index.

No. 11. PHARMACY. Based upon "Remington's Text-book of Pharmacy." By F. E. STEWART, M.D., PH.G., Quiz-Master at Philadelphia College of Pharmacy. Third Edition, Revised.

No. 12. VETERINARY ANATOMY AND PHYSIOLOGY. 29 Illustrations. By WM. R. BALLOU, M.D., Prof. of Equine Anatomy at N. Y. College of Veterinary Surgeons.

No. 13. DENTAL PATHOLOGY AND DENTAL MEDICINE. Containing all the most noteworthy points of interest to the Dental student. By GEO. W. WARREN, D.D.S., Clinical Chief, Penn'a College of Dental Surgery, Philadelphia. Illus.

No. 14. DISEASES OF CHILDREN. By DR. MARCUS P. HATFIELD, Prof. of Diseases of Children, Chicago Medical College. Colored Plate.

Bound in Cloth, $1. Interleaved, for the Addition of Notes, $1.25.

☞ *These books are constantly revised to keep up with the latest teachings and discoveries, so that they contain all the new methods and principles. No series of books are so complete in detail, concise in language, or so well printed and bound. Each one forms a complete set of notes upon the subject under consideration.*

Illustrated Descriptive Circular Free.

JUST PUBLISHED.

GOULD'S NEW
MEDICAL DICTIONARY

COMPACT.

CONCISE.

PRACTICAL.

ACCURATE.

COMPREHENSIVE

UP TO DATE.

It contains Tables of the Arteries, Bacilli, Ganglia, Leucomaïnes, Micrococci, Muscles, Nerves, Plexuses, Ptomaïnes, etc., etc., that will be found of great use to the student.

Small octavo, 520 pages, Half-Dark Leather, . $3.25
With Thumb Index, Half Morocco, marbled edges, 4.25

From J. M. DaCOSTA, M. D., Professor of Practice and Clinical Medicine, Jefferson Medical College, Philadelphia.

"I find it an excellent work, doing credit to the learning and discrimination of the author."

*** **Sample Pages free.**

www.ingramcontent.com/pod-product-compliance
Lightning Source LLC
Chambersburg PA
CBHW020148170426
43199CB00010B/942